About the authors:

Dr. Sukant Khurana is passionate researcher and entrepreneur, exploring neuroscience, drug discovery, biotechnology, artificial intelligence, data science, blockchain, and several areas of emerging technology. He is a well-celebrated artist and author, exploring human senses and human condition. He has also been a tireless fighter for gender equity, social justice, sustainable development, and environment. His earlier grassroots level work caused him several harms, ranging from social media hacking to life-threatening physical attacks. He now focuses on social change through citizen science and data science. He is also a fierce advocate for ethical use of artificial intelligence, social media, and internet of things. He is a famous public speaker, blogger, podcaster, and science and technology advocate, known to communicate in various languages. He has been known for his teaching and coaching skills just as much as research and development work. He is known for lending support and sometimes leading campaigns for raising awareness of mental health, social isolation, and women health in collaboration with various international agencies. Frequently described as one of a kind 21st century polymath, his research, entrepreneurship, and popularization efforts have spanned several branches of natural science, technology, and arts. In recent years, there have been several movies and books on his efforts. He has received most of his advanced scientific training in USA and is currently (2018) based in India. His popular writing spans a large breadth, from very introductory technological texts to cutting edge theoretical works , small scientific monographs to comprehensive reviews of a field, science fiction novels to deep philosophical works, photo-journals to poetry. Like his art, which has been described as not fitting in any one genera, his popular writing also reflects his current interests and emphasis.

Currently he is working in a premier drug-discovery institute, CDRI, which has provided him with a rich exposure to the needs of health-care and how to apply artificial intelligence to solving these problems.

He can be reached at https://twitter.com/sukant_khurana,

https://www.linkedin.com/in/sukant-khurana-755a2343/,

www.brainnart.com,

www.dataisnotjustdata.com,

and https://www.facebook.com/SukantKhuranaauthorsite/

Divyansh Dwivedi

Divyansh Dwivedi is an engineer with bachelor's degree in Computer Science from Lucknow, who is an enthusiastic researcher of Data Science and Artificial Intelligence. Divyansh Dwivedi has worked with Dr. Sukant Khurana for more than a year in Data Science projects and creating popular science Podcasts. He is also a Data Science and AI blogger on Medium. He loves to learn new technologies and is an avid reader.

E-mail-divyanshdwivedi94@gmail.com

LinkedIn https://www.linkedin.com/in/divyansh-d/

Piyush Shrivastava

Piyush Shrivastava has been a research intern working with Dr. Sukant Khurana. He works as a Software Engineer at Boorter and has been working in the field of artificial intelligence for 2 years. As the interest in artificial intelligence rises, there is a need to systematically categorize the tools that will change our future. This has been the inspiration for the book. The hope is that academians and developers across the globe will be able to minimize their efforts in researching the tools and focus on the development instead.

https://www.linkedin.com/in/piyush-shrivastava-13b584a6/

Introduction

The purpose of this compendium is straight-forward. Artificial Intelligence is resulting in fourth industrial revolution. The more the movers and shakers in it, the more it is going to ethical, democratic, and for general good. For a broad participation people should know what are the tools of the trade.

This list is simply a large list of those, current as of June 2018. This book merely serves as a much needed catalogue, with information presented from most popular AI tools websites. This is merely an edited compilation in one place, which was needed for the field but missing. We hope to follow it up with more detailed systematic analysis of all AI tools soon.

Dedication

This book is dedicated to the vision of creating a smarter and more equitable world through artificial intelligence. It is also dedicated to the goal of making India, the current home of all three authors, into an artificial intelligence hub of global consequences.

Deep Learning

PyTorch

PyTorch based on python library is built to provide a versatile deep learning construction platform. The workflow of PyTorch is very close experience as directly using python's scientific computing library like numpy. Some other advantages of using PyTorch are its multi-GPU support, custom data loaders, and simplified preprocessors. PyTorch uses an imperative paradigm, that is, each line of code required to build a graph, defines a component of that graph. On these parts, we can perform computations independently itself.

Type: Library

By Adam Paszke, Sam Gross, Soumith Chintala, Gregory Chanan in October 2016.

GUI Supported: No

Features:

- PyTorch now fully supports advanced indexing; following numpy has advanced indexing rules. Also added FFT (Fast Fourier transform).
- PyTorch provides a system for users where they can write their own C++ / CUDA extensions.
- An n-dimensional Tensor, similar to numpy but can run on GPUs.
- Automatic differentiation for building and training neural networks.

Comparison:

- PyTorch is defined at runtime you can use our favorite Python debugging tools such as pdb, ipdb, PyCharm debugger or old trusty print statements as compared to Tensor flow.
- PyTorch supports declarative data parallelism to wrap any module and it will be parallelized over batch dimension.

Access/Source: Closed Access/Open Source.

License: BSD 3-Clause

Language: Python, CUDA, C++.

API Supported: Yes

Total Users/Projects: https://github.com/pytorch

Platform: Windows, Linux, Unix.

Community: https://discuss.pytorch.org/latest

Videos: https://www.youtube.com/watch?v=nbJ-2G2GXL0

Reviews:https://www.reddit.com/r/MachineLearning/comments/5w3q74/d_so_pytorch_vs_tensorflow_whats_the_verdict_on/

Website: https://pytorch.org/

TfLearn

TFlearn is a modular and easy-to-use deep learning library created on top of Tensorflow. It was intended to implement a higher-level API to TensorFlow for facilitating and speeding-up experiments, while remaining fully open and compatible with it. The TFlearn is easy to understand and easy to use high-level API for implementing deep neural networks. This library is a powerful helper function to train any TensorFlow graph, with the support of multiple inputs, outputs, and optimizers. The library API supports and works in most of the recent and famous deep learning models such as ConvNet, LSTM, Batch Normalization, Activation Layers, Residual networks, Generative adversarial networks, etc.

Type: Library

Founded in 2016.

GUI Supported: No

Features:

- Easy-to-use and understand high-level API for implementing deep neural networks, with tutorial and examples.
- Fast prototyping through highly modular built-in neural network layers, regularizes, optimizers, metrics.
- Full transparency over Tensorflow. All functions are built over tensors and can be used independently of TFLearn.
- Powerful helper functions to train any TensorFlow graph, with support of multiple inputs, outputs, and optimizers.
- Easy and appealing graph visualization, with details about weights, gradients, activations, and more.
- Effortless device placement for using multiple CPU/GPU.

Comparison:

- Its API is closer to that of TensorFlow as compared to Keras.
- It offers slightly better performance than Keras.
- It allows one to use Python arrays directly. Keras needs NumPy arrays as compared to Keras.

Access/Source: Closed Access/Open Source.

License: MIT

Language: Python

API Supported: http://tflearn.org/doc_index/#API

Total Users/Projects: https://github.com/tflearn/tflearn

Platform: Windows, Linux.

Community: No

Videos: https://www.youtube.com/watch?v=pFbyPWLB17I

Reviews: https://www.predictiveanalyticstoday.com/tflearn/

Website: http://tflearn.org/

Swift AI

Swift AI is a currently a small effort, a high-performance library written in a swift language, which includes machine learning and artificial intelligence libraries. Swift can support genetic algorithms, matrix computations, digital signal processing, artificial neural networks and more. You can use these tools to integrate Artificial Intelligence into your apps quickly.

Type: Library

By Collin Hundley in December 1, 2016.

GUI Supported: No

Features:

- Feed-Forward Neural Network
- Fast Matrix Library
- High-performance AI library
- Written entirely in Swift

Comparison: No

Access/Source: Closed Access/Open Source

License: MIT

Language: Swift

API Supported: No

Platform: Mac, IPhone, IPad

Community: No

Videos: https://www.youtube.com/watch?v=Xw-Ja6uXwNI

Reviews: https://www.producthunt.com/posts/swift-ai

Website: https://github.com/Swift-AI/Swift-AI

Chainer

Chainer is a standalone open-source framework, which based on Python for deep learning models. Chainer provides a flexible, intuitive, and high-performance means of implementing a full range of DL models, including state-of-the-art examples such as recurrent neural networks and variational autoencoders. It provides a flexible Python-based framework which is used for quickly and intuitively writing complex neural network architectures. Chainer provides an easy to use platform for training multi-GPU instances. It also automatically logs results, graph loss and accuracy, and produces output for visualization.

Type: Framework

By Toru Nishikawa and Daisuke Okanohara in June 2015.

GUI Supported: No

Features:

- Chainer supports CUDA computation.
- Chainer supports various network architectures including feed-forward nets, convnets, recurrent nets and recursive nets.
- Chainer is faster, Scalable and easy to use framework.
- Chainer is a flexible and intuitive framework for Neural Networks.
- Parallel and Automatic Optimization.

Comparison:

- Better GPU & GPU data center performance than Tensor Flow.
- It is highly flexible and easiness to write neural networks as compared to other deep learning frameworks.
- CuPy also makes it easy to write custom kernels as compared to keras.

Access/Source: Open Access/Open Source

License: MIT

Language: Python

API Supported: https://docs.chainer.org/en/stable/compatibility.html

Platform: Windows, Linux

Community: https://github.com/chainer-community

Videos: https://www.youtube.com/watch?v=O3bZutlmqv4

Reviews: No

Website: https://chainer.org/

Deeplearninig4j

Eclipse Deeplearning4j is an open-source distributed-deep learning library, relying on the widely used programming language, Java and Scala. Integrated with Hadoop and Spark and powered by the open-source numerical computing library ND4J, it operates with both central processing and graphics processing units.

Deeplearning4j has been utilized as a part of many academic and business applications. Its framework allows neural nets, such as restricted Boltzmann machines and recurrent nets to be added to one another to create deep nets of varying types. It also claims to provide a matrix library much faster than numpy.

Type: Library

Founded in 2000.

GUI Supported: No

Access/Source: Closed Access/Open Sourced

License: Apache 2.0

Language: Java, Scala

API Supported: Yes

Platform: Linux, Mac, Windows, Android

Community: https://deeplearning4j.org/devguide

Video: https://www.youtube.com/watch?v=N5sQcOOtehY

Review: https://www.g2crowd.com/products/deeplearning4j/reviews

Website: https://deeplearning4j.org/

Repository: https://github.com/deeplearning4j/deeplearning4j

CNTK

The Computational Network Toolkit is a C++ library produced by Microsoft. Its prominent use is to manipulate computational networks and has been released under a permissive license. Commercially available, it is widely used for language processing. Execution from multiple machines is possible and multiple GPUs can be used to accelerate learning.

CNTK 2.0 has a Python API. The library is compatible with the Python IDE Anaconda and can be used with computational neural network for speech analysis.

Type: Toolkit

Established in January, 2016

GUI Supported: No

Access: Open-sourced

License: MIT

Language: C++

API Supported: Yes

Platform: Linux, Windows

Community:https://docs.microsoft.com/en-us/cognitive-toolkit/contributing-to-cntk

Video: https://www.youtube.com/watch?v=4-1btPONlCM

Review: http://minimaxir.com/2017/06/keras-cntk/

Website: https://www.microsoft.com/en-us/cognitive-toolkit/

Repository: https://github.com/Microsoft/CNTK

Caffe

Caffe is a C++ and Python-based deep learning framework, developed by Berkeley AI Research (BAIR). Caffe is released under the BSD 2-Clause license. Although modelling capability for Recurrent Neural Networks is lacking, it's extensible design, fast execution, quick deployment on a standard model and broad range of community contributors makes it a popular choice for academic research projects and industrial applications.

Data in Caffe in stored in blobs, a 4-dimensional data structure with on-demand memory allocation. The code for the library is modular and based on abstraction, allowing easy extensions and extensive testing.

Type: Library

Released in December, 2013

GUI Supported: No

Access: Open-sourced

License: BSD

Language: C++

API Supported: Yes

Platform: Linux, Windows, macOS

Community: https://groups.google.com/forum/#!forum/caffe-users

Video: https://www.youtube.com/watch?v=rvMVqPsXL10

Review: https://www.g2crowd.com/products/caffe/reviews

Website: http://caffe.berkeleyvision.org/

Repository: https://github.com/BVLC/caffe

SystemML

If you need an adaptable machine learning system that can automatically scale to Spark and Hadoop clusters, Apache SystemML could be your choice. Aside from providing optimal workplace for using big data, it can run on top of Apache Spark to automatically scale your data line by line, examining whether your code should be run on the driver or an Apache Spark cluster. Currently, SystemML lacks additional deep learning with GPU capabilities such as importing and running neural network, but they are likely to appear in future builds.

Type: Library

Foundation: November, 2015

GUI Supported: No

Access: Open-sourced

License: Apache 2.0

Language: Java

API Supported: Yes

Platform: Linux, Windows, macOS

Community: http://systemml.apache.org/community

Video: https://www.youtube.com/watch?v=5Y2k1aPqW6g

Website: http://systemml.apache.org/

Repository: https://github.com/apache/systemml

Torch

Torch is a scientific computing framework written in the programming language Lua with wide support for machine learning algorithms that puts GPUs first. It comes with efficiency and easiness in use, thanks to LuaJIT, an easy and fast scripting language, under which any library in C or C++ can become a Lua library.

Torch aims at maximum flexibility and minimizing computation time along with making the process simple for the user. You can build complex neural network topologies, and parallelize them in an efficient manner over CPUs and GPUs. It has embeddable ports to both iOS and Android. NYU, Facebook AI lab and Google Deepmind are its prominent users in the industry.

Type: Library

Foundation: October, 2002

GUI Supported: No

Access: Open-sourced

License: BSD

Language: Lua, LuaJIT, C, CUDA, C++

API Supported: Yes

Platform: Linux, Windows, Mac OS X, iOS, Android

Community: http://torch.ch/support.html

Video: https://www.youtube.com/watch?v=L1sHcj3qDNc

Review: https://www.g2crowd.com/products/torch/reviews

Website: http://torch.ch/

Repository: https://github.com/torch

Apache SINGA

Apache SINGA is an Apache project going through incubation for developing an open source machine learning library, supported by Apache Incubator. While maintaining a flexible architecture for scalable distributed machine learning, it can also be run over a varied range of software.

The project's software stack consists of three major components – core, IO and model. Although the project has yet to be endorsed by The Apache Software Foundation, with Tensor abstraction, SINGA would be able to run a wide range of learning models. SINGA mainly focuses on health-care applications, and its trial is available to download for free.

Type: Library

Released in October, 2015

GUI Supported: No

Access: Open-source

License: Apache 2.0

Language: C++, Python, Java

API Supported: Yes

Platform: Linux, Windows, macOS

Community: http://singa.apache.org/en/develop/how-contribute.html

Video: https://www.youtube.com/watch?v=ZN8dJHgSi84

Website: http://singa.apache.org/

Repository: https://github.com/apache/incubator-singa

Keras

Keras is written in Python, the most popular language used by developers for machine learning. It is an open-source neural network library that is capable of running on multiple other machine learning frameworks, such as Deeplearning4j, Tensorflow, and Theano.

Its key features are minimalism, easy extensibility and modularity. Keras was initially designed to be interface, keeping quick experimentation in mind. But it has recently gained popularity to become the second-fastest growing deep learning framework after Google's Tensorflow.

Type: Library

Released in March, 2015

GUI Supported: No

Access: Open-sourced

License: MIT

Language: Python

API Supported: Yes

Platform: Linux, Windows, macOS

Community: https://groups.google.com/forum/#!forum/keras-users

Video: https://www.youtube.com/watch?v=L1sHcj3qDNc

Review: https://www.capterra.com/p/171047/Keras/

Website: https://keras.io/

Repository: https://github.com/keras-team/keras

Edward

In recent years, there has been a rise in the use of probabilistic programming. Edward is one such Python-based library, designed for probabilistic modeling, inference, and criticism. Built on top of Tensorflow, it combines the fields of machine learning, Bayesian statistics, and probabilistic programming. It acts as a black box tool for fast experimentation and research with probabilistic models, ranging from classical hierarchical models on small data sets to complex deep probabilistic models on large data sets.

Type: Library

Released in 2016

GUI Supported: No

Access: Open-sourced

License: Apache 2.0

Language: Python

API Supported: Yes

Platform: Linux, Windows, macOS

Community: http://edwardlib.org/community

Video: https://www.youtube.com/watch?v=PvyVahNl8H8

Website: http://edwardlib.org/

Repository: https://github.com/blei-lab/edward

Theano

Another Python-based library favored by developers is Theano, specifically used for deep learning. Primarily developed by a machine learning group, it was later open-sourced. It offers the flexibility to define and evaluate mathematical expressions.

Theano has been able to rival C implementations in terms of speed. Some models based on it have also surpassed C by taking advantage of the GPUs. While providing stability optimization, it also offers dynamic generation of C code, so that expressions can be evaluated faster. Theano also includes an interface to the compiler, combining the best of scientific libraries and a high-level language.

Type: Library

Foundation: March, 2015

GUI Supported: No

Access: Open-sourced

License: BSD

Language: Python

API Supported: Yes

Platform: Linux, Windows, macOS

Community: https://groups.google.com/forum/#!forum/theano-users

Video: https://www.youtube.com/watch?v=BuIsI-YHzj8

Review: https://www.g2crowd.com/products/theano/reviews

Website: http://deeplearning.net/software/theano/

Repository: https://github.com/Theano/Theano

MXNet

A flexible and modern library for deploying deep neural networks is Apache MXNet. It is highly scalable, and provides rich APIs for multiple languages like Python, Java, and MATLAB. With MXNet, it's easy to postulate the location of each dataset. It also automates many complex calculations by emphasizing on speeding up the development and execution of large-scale models.

The MXNet library is released under Apache License 2.0, and the code is freely available on GitHub. Being portable and distributed on cloud, it is supported by may Cloud Providers like Intel, Microsoft and Baidu.

Type: Library

Foundation: March, 2015

GUI Supported: No

Access: Open-sourced

License: Apache 2.0

Language: C++, Python, R, Julia, JavaScript, Scala, Go, Perl

API Supported: Yes

Platform: Linux, Windows, macOS

Community: https://groups.google.com/forum/#!forum/theano-users

Video: https://www.youtube.com/watch?v=iLqEhCqCQ70

Website: https://mxnet.apache.org/

Repository: https://github.com/apache/incubator-mxnet

ND4J

A popular choice with production environments, ND4J is designed to run proficiently with minimum RAM requirements. It's written in C++, operates on Java Virtual Machine (JVM) and has a limited compatibility with Java, Scala and Clojure.

ND4J is open-sourced and released under Apache License 2.0. It can perform complex computations in integration with Apache Hadoop and Sparkto in production environments. It is popular in places requiring simulations with heavy computational requirements.

Type: Library

Released in September, 2014

GUI Supported: No

Access: Open-sourced

License: Apache 2.0

Language: Java, C++

API Supported: Yes

Platform: Linux, Windows, macOS, Android

Community: https://gitter.im/deeplearning4j/deeplearning4j

Video: https://www.youtube.com/watch?v=ciSipH4-hnw

Website: http://nd4j.org/

Repository: https://github.com/deeplearning4j/nd4j

NeuroSolutions

NeuroSolutions is a visual programming environment for neural network simulation. Different visual components are available to design, test and analyse the neural networks. The network structured available are currently limited to feed-forward, recurrent, radial-basis function, and Kohonen maps. Designed by NeuroDimension Inc., there are three proprietary licenses available - Pro Single-user, Single-user and Student Single-user. NeuroSolutions also avails speed improvements through parallel computing. NeuroSolutions Accelerator harnesses the massive power of GPUs and multi-core processors to discount the training time from hours to possibly even minutes.

Type: Software

Foundation: March, 2015

GUI Supported: Yes

Access: Paid

License: EULA

API Supported: Yes

Platform: Windows

Community: https://groups.google.com/forum/#!forum/theano-users

Video: https://www.youtube.com/watch?v=Dn8tC-5Am2M

Review: http://www.neurosolutions.com/resources/ieee.html

Website: http://www.neurosolutions.com/

Neural Designer

Unlike many other commercial softwares, Neural Designer is a desktop-based application, which utilizes neural networks to gather intelligence. It has been developed by the company Artelnics, based on the open-source library OpenNN. Visualisation with high performance computing, availability for all size of businesses and ease of use makes it a highly popular choice. Although the software is proprietary with desktop and cloud versions available, there is also a free version with limited features.

The three basic components of Neural Designer are Neural Editor, Neural Viewer, and Neural Engine. The computational tasks are executed by neural engine in the background. There is also a provision of pre-analysis tools, exporting network equations and comprehensively customisable reports.

Type: Software

Foundation: 2015

GUI Supported: Yes

Access: Paid

License: Proprietary Software

Language: C++

API Supported: Yes

Platform: Linux, Windows, OS X

Video: https://www.youtube.com/watch?v=9hKHVEm16jU

Review: https://www.g2crowd.com/products/theano/reviews

Website: https://www.neuraldesigner.com/

ConvNetJS

ConvNetJS provides an open source library, which is based on JavaScript for training Deep Learning models such as Neural Networks entirely in a browser. Open a tab and start training. No software requirements, no compilers, no installations, no GPUs. It is simple to adopt and fast without using a GPU, which implies it works almost everywhere. Andrej Karpathy creates this library, a PhD Student at Stanford, to develop neural networks more available. No GPUs are managed to make things go faster, but the raw JavaScript seems to do the job very well for beginners and mid-level users.

PaddlePaddle

The Paddlepaddle is an easy-to-use deep learning platform, which provides an efficient, adjustable and scalable, which was initially developed at Baidu company for applying deep learning models to Baidu products. PaddlePaddle is created to be independent of computing foundation. It can run on top of Apache Hadoop, Spark, Mesos, and others apache tools. It has a keen interest in Kubernetes because of its flexibility, efficiency, and rich features. It uses the distributed environment to speed up the forward and backwards passes through parallel processing. It is aimed towards the production environment.

FANN

Fast Artificial Neural Network is a platform, which provided an open source library for the neural network, and in this platform, the C language is used in multilayer artificial neural networks with the support for both fully connected and not entirely connected channels. Cross-platform execution in supports both fixed and floating point. It also provides a framework for the simple handling of training data, and it is an easy-to-use framework, which is also versatile, well documented, and fast. An Artificial Neural Network (ANN) is a simplified emulation of one part of our brains. Precisely, they simulate the activity of neurons within the mind. This technique falls under the field of Artificial Intelligence.

Neuroph

Neuroph is a Java neural networks framework, which is a lightweight framework for developing joint neural networks designs. It provides a Java neural network library as well as a GUI tool that supports creating, training and saving neural networks. Neuroph released as open source under the Apache 2.0 license. Neuroph is an object-oriented neural network structure. It can be managed to create and train neural networks in Java programs. Neuroph provides Java class library as well as GUI tools simple Neurons for building and preparing neural networks. It contains a small number of basic classes, which corresponds to essential neural network concepts.

Predictive Analytics

(Business intelligence tools, involving both deterministic and AI components)

Dundas BI

Dundas BI is a platform, which is used for business intelligence like building interactive reports, scorecards, dashboards, and use of other tools. It is entirely customizable with business intelligence tools and data visualization software. It comes with simple self-service, low experience need, powerful visualizations, dashboards, and reporting. The users now have real-time data analytics and result from any data source, on any device, all delivered on one flexible, open platform. Dundas data visualization helps organizations to improve efficiency.

Type: Software

Founded in 1992

GUI Supported: Yes

Features:

- Easy, smart drag and drop design tools.
- Fully customizable visualizations to create personal and professional dashboard views.
- Wide range of visualizations, styles and themes.
- Responsive design options.
- Navigation (drill-down to another dashboard or view, forward/back buttons)
- Data filters (dates, categories, etc.)
- Customizable buttons.
- Customize how the dashboard responds to interactions (mouse-clicks, hovers, etc.)
- Supports ESRI shape files that are built in or provided by the user.
- Visualizes data via color-coded shapes ("choropleth") and in addition, predefined map element properties can bind to data (i.e. color, border, and other visual properties).
- Visualizes data via bubbles or symbols of varying sizes centered above geographic areas.

Comparison:

- Dundas BI has an extensive API that allows for extensive flexibility. The available APIs include .NET, JavaScript, and REST APIs as compared to Microsoft BI.
- Load balancing is included right out of the box with Dundas BI.
- Dundas BI has outstanding mobile support out of the box.
- Dundas BI represents a comprehensive BI platform in a single package.

Access/Source: Closed Access/Closed Source.

License: Paid

API Supported: Yes

Platform: Windows, Mac

Community: https://www.dundas.com/support/

Videos: https://www.youtube.com/watch?v=0BD3Edhfnso

Reviews: https://www.softwareadvice.com/bi/dundas-bi-profile/

Website: https://www.dundas.com/

Silvon

Silvon also provides a platform, which is used for business intelligence, operational planning, and reporting solutions for companies and merchants of consumer goods and others. Using pre-built analytics, reports and scorecards, the software provides a depiction of performance across sales, marketing, and operational areas. Silvon addresses critical operational areas of manufacturing and distribution.

Type: Software

Foundation: In 1987 in United States.

GUI Supported: Yes

Features:

- Ad hoc Analysis
- Ad hoc Query
- Ad Hoc Reports
- Key Performance Indicators
- OLAP
- Performance Metrics
- Budgeting & Forecasting
- Dashboard
- Data Analysis
- Data Visualization
- Profitability Analysis
- Strategic Planning
- Trend / Problem Indicators

Comparison:

- Ease of implementation and use as compared to several other software.
- Cost effective solution.
- Ease of use in creating views and downloading to Excel.

Access/Source: Closed Access/ Closed Source

License: Paid

API Supported: No

Platform: Windows

Community: https://forum.silvon.com/

Videos: https://www.youtube.com/watch?v=XGt07AkiuVg

Reviews: https://www.softwareadvice.com/bi/stratum-profile/

Website: https://www.silvon.com/

Woopra

Woopra is an analytics platform, which provides the support to understand client's behavior across various devices and touch points, including help desk, email, and chat on website as well as the mobile app. It utilizes this data to develop complete forms for every user, synchronizing data from many sources, continuously following activity on web and mobile and automatically updating patterns in real time. This analytical service can also create customer segments, such that when a user behavior changes it automatically update the customer segment. You can create custom live dashboards based on your core KPIs. One can use several third-party apps, which integrate to automate triggered actions in other apps.

Type: Software

By Elie Khoury, Jad Younan in March 1, 2012.

GUI Supported: Yes

Features:

- Third party app integration through AppConnect
- Automatic real-time segment updating
- Build customer segments
- Creates custom reports
- Data from email, live chat, and help desk
- Deliver triggered actions
- Funnel reports
- Integration with Salesforce, Marketo, Zendesk and several other
- Live KPI dashboards
- Real-time analytics and data
- Real-time notifications

Comparison:

- It supports many integration like Salesforce, Eloqua, Marketo, BlueKai, Zendesk, Optimizely, Freshdesk, Box, Dropbox, Google Drive, Stripe, Magento, WooCommerce, and WordPress, while IBM Marketing Cloud does not.

- It supports real-time notifications, scheduled tasks, track all customer activity, and track new and unidentified web and mobile users, trigger JavaScript or Webhooks actions, while most other softwares don't.

Access/Source: Closed Access/Closed Source

License: Paid

API Supported: https://docs.woopra.com/v3.0/reference

Platform: Windows, Linux, Android, Mac, Web Browsers.

Community: https://www.yireo.com/forum/woopra

Videos: https://www.youtube.com/watch?v=Su9MVAU5FzQ

Reviews: https://www.g2crowd.com/products/woopra/reviews

Website: https://www.woopra.com/

Birst

Birst is a platform for advanced networked business analytics. Organizations can gain a new level of trusted insight and decision making by joining their data via a network of analytics services. It scales from individuals to the enterprise in a way that it turns to be smart, connected, and scalable. Birst is a web-based networked BI and analytics solution that combines insights from various teams and helps in making informed decisions. It also offers a unified semantic layer that maintains common definitions and key metrics. The tool enables decentralized users to augment the enterprise data model virtually without compromising data governance.

Type: Software

Foundation: In 2004 in San Francisco, CA.

GUI Supported: Yes

Features:

- Adaptive User Experience
- User Data Tier
- Blends centralized and decentralized data
- Multi-tenant Cloud Architecture
- Enterprise Analytics
- Embedded Analytics
- Automated Data Refinement (ADR)
- Infinite Connectivity Framework
- Responsive and integrated HTML5 dashboards and visual discovery interfaces

Comparison:

- It supports integration with the popular softwares like Salesforce, Snap Logic, Tableau, NetSuite, Zoopla, MarcomCentral, while other softwares don't.
- It helps create, schedule, alert, and deliver highly formatted reports, and provide predictive analytics engine with easy deployment as compared with others, which do not.

Access/Source: Closed Access/Closed Source

License: Paid

API Supported: Yes

Platform: Mac, IPhone/IPad, Web Browsers

Community: https://www.birst.com/services/support/

Videos: https://www.youtube.com/watch?v=adEHF5vzOjU

Reviews: https://www.g2crowd.com/products/birst/reviews

Website: https://www.birst.com/

Oracle Hyperion

Oracle Hyperion is a web-based platform, which provides broadcasting and analysis, and financial association in a single, highly efficient software solution. Oracle Hyperion Financial Management, which is the part of Oracle EPM Suite, validates several things, such as collection, integration and reporting of financial outcomes in numerous GAAP as well as IFRS, and the settlement of changes between the multiple standards. Hyperion Financial Management can be organized instantly to incorporate data from several Oracle and non-Oracle transactional systems, provide topside reporting to IFRS requirements, and support organizations make the transition to IFRS.

Type: Software

Founded in 1998 and acquired by Oracle in 2007.

GUI Supported: Yes

Features:

- Flexible custom dimensions allows users to manage members in dimensions distinctly by plan.
- New task list types, such as Copy Version and Job Console, allows for easier navigation through the system.
- Ability to validate target members from mapping tables.
- Dynamic Scripting - user can choose between traditional formula expressions using DRM native functions, or developing java script formulas for derived properties and validation rules in DRM.
- Users now have the ability to format annotation text.
- FDMEE now supports loading ICP transactions into ICT Module.

Access/Source: Closed Access/Closed Source

License: Paid

API Supported:
https://docs.oracle.com/cd/E17236_01/epm.1112/drm_api.html

Platform: Windows, Web Based.

Community:
http://www.oracle.com/technetwork/middleware/performance-management/community/index.html

Videos: https://www.youtube.com/watch?v=7lzFVqU_Ojw

Reviews: https://www.g2crowd.com/products/oracle-hyperion-planning/reviews

Website: https://www.oracle.com/applications/performance-management/products/business-planning/hyperion-planning/index.html?utm_source=PredictiveAnalyticsToday&utm_medium=Review&utm_campaign=PAT

Yellowfin

Yellowfin is a BI software organization, founded in Melbourne in 2003 that set out to improve the general BI program because the originators felt that common BI had become too complicated and costly. Yellowfin's strength lies in selling its software to OEM groups, where its BI functionality mixed into other vendors' products and applications. Yellowfin is a platform which provides mature analytics, and it is user-friendly. It has evolved from a reporting and dashboard to support an emerging style of Business Intelligence characterized by governed data discovery and collaboration.

Type: Software

Foundation: By Glen Rabie and Justin Hewitt in 2003.

GUI Supported: Yes

Features:

- Reports interface for standard and self-service reports is intuitive and easy to use.
- Offers a variety of attractive graph and chart formats.
- Using formulas based on existing data elements; users can create and calculate new field values.
- Users can drill down and explore data to discover new insights.
- Business users have the ability to filter data in a report based on predefined or auto-modeled parameters.
- Analyze current and historical trends to make predictions about future events.
- It Supports Mobile BI, Collaborative BI, Mapping, Storyboard.

Comparison:

- Yellowfin is fully embeddable and makes it easy for the users to seamlessly embed excellent BI and analytics into their application, while most other similar products do not.

Access/Source: Closed Access/Closed Source

License: Paid

API Supported: Yes

Platform: Windows, Mac, Web Browsers.

Community: https://community.yellowfinbi.com/

Videos: https://www.youtube.com/watch?v=Bsn0zdtsvdc

Reviews: https://www.softwareadvice.com/bi/yellowfin-profile/

Website: https://www.yellowfinbi.com/

BIME

BIME is a cloud-based platform, which provides Business Intelligence (BI). Zendesk was the first organization to provide front-end BI skills for Google Big Query. It was also the first BI app accessible in Google Store. BIME has collaborated with Google to offer robust significant data analytics experience. BIME has partnerships with technology companies like Google, Amazon, Appirio, Noovle, and more. BIME is a simple yet excellent service that connects to and analyses data in any organization. It is a simple-to-use BI app based on cloud computing innovations and data visualization.

Type: Software

By Zendesk in 2009.

GUI Supported: Yes

Features:

- 16 advanced visualizations, with 500 customization points.
- Advanced capability to build dashboards on all devices.
- Automatic chart selection based on best practices in Dataviz.
- Complex calculated models and "what if" analysis.
- Connectors to Big Data sources (Big Query, HANA, Redshift).
- Data storytelling and discovery with bookmarks and slides.
- Fast, distributed and parallel column-store in-memory engine.

Comparison:

- It Supports Web and social media analytics: Google Analytics, Twitter, Facebook, and YouTube as compared to Kaseya VSA.
- On the fly mixing of data sources through the Query Blender.
- It supports SQL performance boost – queries optimized for each database.
- Native connectors to 35+ on premise and online data sources as compared to other software.

Access/Source: Closed Access/Closed Source.

License: Paid

API Supported: Yes

Platform: Windows, Mac, Web Based, Windows Mobile.

Community: https://bime.zendesk.com/hc/en-us/community/topics

Videos: https://bime.zendesk.com/hc/en-us/articles/217604948-Introducing-BIME

Reviews: https://www.g2crowd.com/products/bime-analytics/reviews

Website: https://www.bimeanalytics.com/

Google Analytics

Google Analytics is web analytics, which is a free tool provided by Google that assists users to measure website traffic and collect vital data regarding their website visitors. To measure and improve your website, Google Analytics is one of the best free of cost tools. It can be used for several other web search analytics. Given that it is so widely used and discussed we are not going into much details of it here.

Type: Web Analytics

Released on November 14, 2005.

GUI Supported: Yes

Features:

- Advertising Reports
- Campaign Measurement
- Cost Data Import
- Mobile Ads Measurement
- Remarketing
- Search Engine Optimization
- Advanced Segments
- Annotations
- Content Experiments
- Dashboards
- Custom Reports
- Real-Time Reporting
- Audience Data & Reporting
- Browser / OS
- Custom Dimensions
- Flow Visualization

Comparison:

- It supports Google Play Integration, iOS and Android SDKs, product integrations, attribution model comparison tool, data-driven contribution, ecommerce reporting, goals/goal flow as compared to Dundas BI software.

- It also supports multi-channel funnels, alerts, intelligence events, event tracking, in-page analytics, site search, and site's speed analysis as compared, while most softwares don't.

Access/Source: Closed Access/Closed Source.

License: Free as long as you do not exceed 5 million impressions per month.

API Supported: https://developers.google.com/analytics/

Platform: Windows, Linux, Mac, Android, IPhone/IPad, Web Based.

Community: https://developers.google.com/analytics/community/

Videos: https://www.youtube.com/watch?v=lZf3YYkIg8w

Reviews: https://www.g2crowd.com/products/google-analytics/reviews

Website: https://www.google.com/analytics/#?modal_active=none

Open Web Analytics

Open Web Analytics (OWA) is a web analytics software, which is open source, you can use to track and analyses that how people use their websites and applications. OWA is licensed under GPL and it provides website owners and developers with easy ways to add web analytics to their sites using pure JavaScript, PHP, or REST based APIs. OWA build with popular content management frameworks and came with built-in support for tracking websites. It includes tracking of users, roles, content and in-page user behavior and uses OWA Data Access API to show reports and stats inside Drupal. This analytics comes with built-in support for tracking a user's full-page experience including mouse movements, scrolling, and key-presses includes standard tracking features.

Type: Web Analytics

Released on September 27, 2017.

GUI Supported: Yes

Features:

- Open Web Analytics tracks where on a page, and on what elements, visitors click.
- It provides heat maps that shows where on a page visitors interact the most.
- OWA has a wordPress plugin and can integrate with other plugins like media wiki.

Comparison:

- It provides click heat map and mouse movement as compared to other software, which don't provide this kind of features.
- It support Command Line Interface instead of GUI to perform admin tasks as compared to google analytics which don't provide.

Access/Source: Closed Access/Open Source.

License: GPL v2

Language: PHP

API Supported: Yes

Platform: Windows, Linux, Mac, Web Based.

Community: http://www.openwebanalytics.com/?cat=12

Videos: https://www.youtube.com/watch?v=gg8ay8M9KnY

Reviews: http://softwareproductreview.com/open-web-analytics-review/

Website: http://www.openwebanalytics.com/

Piwik

Matomo, which is formerly known as Piwik, is an open analytics platform currently used by people, organizations and authorities all over the globe. By matomo, your data will always be yours. It is an open source software program, which is freely available for Web analytics. The program gives detailed reports on your website traffic, including famous statements such as keywords visitors and the search engines used to locate your page, the language they speak, your favorite sheets and additional analytical records. It is a PHP and MySQL software program that every web publishers need to download and install on your web server. After the initial set-up, a JavaScript code provided, that you must copy and paste it on any websites you want to trace using the Web analytics software.

Type: Web Analytics

Released on March 28, 2018.

GUI Supported: Yes

Features:

- Real Time Data Updates.
- Customizable Dashboard
- All Websites Dashboard
- Row Evolution
- Analytics for Ecommerce
- Goal conversion tracking
- Event Tracking
- Content Tracking
- Site Search Analysis

Comparison:

- The key to using data insights is the ability to manipulate the dataset.
- Creating custom data sets for tiles/tabs is straightforward as compared to other analytic software, which don't provide this feature.
- It's free, community supported with constant upgrades. It is compatible with most websites and provides awesome statistics.

Access/Source: Closed Access/Open Source.

License: GNU GPL v3

Language: PHP

API Supported: https://developer.matomo.org/api-reference

Platform: Web Based, Windows, Mac, Linux, IPhone/IPad

Community: https://developer.matomo.org/support

Videos: https://www.youtube.com/watch?v=yA2NUur0770

Reviews: https://www.g2crowd.com/products/piwik/reviews

Website: https://matomo.org/

Adobe Analytics

Adobe Analytics, which is formerly known as Omniture Site Catalyst, provides an analytics platform for the solution to implementing real-time analytics and precise segmentation across all marketing channels. The platform helps clients deliver better customer experiences by providing real-time insight into the performance of marketing initiatives. As a conclusion, clients are better able to refine and tailor personalized marketing campaigns and digital experiences across every channel. The Adobe Analytics platform helps to build a holistic view of marketing activities by changing customer communications into insights. A user can control and share real-time information which helps in identify issues, discover opportunities and measure by interactive and straightforward dashboards and records.

Type: Web Analytics

Founded by Adobe Systems in October 24, 2012.

GUI Supported: Yes

Features:

- Accurately measure on total web traffic and present data elegantly.
- Configure and accurately measure an appropriate measure of engagement, such as average session length, page views per session, amount of scrolling, etc.
- Accurately determine where most people enter your site and where they leave and present data elegantly.
- Able to set up custom variables to track any user activity on the site.
- Report on user activity in the past and set up predictive models for the future.
- Break down a population of users by demographics, sequence actions, time, and able to build custom segments.
- Analytic tools are also available for mobile and tablet versions of websites
- Able to access user-specific data such as location, language, gender, and activity while on the site.

Comparison:

- Adobe Analytics Cloud allows you to closely track your digital marketing campaigns and their performances, recognize and create different types of data segments for remarketing as compared to other analytics software, which don't provide this feature.
- It offer a customized shopping experience to your site visitors, and works with bare-minimum data latency, particularly during the periods of peak traffic activity and lets you do a lot more.
- Analytics Cloud can also connect to a wide range of native and 3rd party tools to easily bring in 3rdparty data to the tool.

Access/Source: Closed Access/Closed Source.

License: Paid

API Supported: https://www.adobe.io/apis/experiencecloud/analytics.html

Platform: Windows, Linux, Mac.

Community: https://forums.adobe.com/community/experience-cloud/analytics-cloud/analytics

Videos: https://www.youtube.com/watch?v=IkbW2MywpXE

Reviews: https://www.g2crowd.com/products/adobe-analytics/reviews

Website: https://www.adobe.com/data-analytics-cloud/analytics.html

DataRobot

DataRobot provides a platform for data scientists of all experience levels to build and deploy specific predictive models in machine learning in a short time. It allow the users to create and implement highly accurate machine learning models in a portion of the time. The technology addresses the significant deficiency of data scientists by improving the speed and economics of predictive analytics. This platform does massively parallel processing to train and estimate 1000's of models in Python, MLlib, R, H2O and other libraries. It includes hundreds of open source machine learning algorithms to train the machine learning models.

Type: Web Analytics

Founded by Jeremy Achin, Thomas DeGodoy in June 1, 2012.

GUI Supported: Yes

Features:

- Drag and drop the datasets from any location.
- Automatically built and evaluate 100's of machine learning models om a short time.
- Make predictions and get insight graph of the data.
- Collaboration & Social Business Intelligence integration.

Comparison:

- Ease of use and deploy as administer as compared to IBM Watson.
- Better self-contained extraction, transformation & loading (ETL) & data storage capabilities as compared to IBM Watson, which don't have this feature.
- Easy to use self-service data preparation feature as compared to Tableau.

Access/Source: Closed Access/Closed Source

License: Paid

Language: Python

API-Supported:https://datarobot-public-api-client.readthedocs-hosted.com/en/v2.9.0/api/index.html

Platform: Web Based, Windows, Mac

Community: https://blog.datarobot.com/

Videos: https://www.youtube.com/watch?v=_JOKlSwKJ10

Reviews: https://www.predictiveanalyticstoday.com/datarobot/

Website: https://www.datarobot.com/

BigML

BigML is a robust machine learning service that offers an easy-to-use interface to get predictions out of it by imports your data. The service is so easy to use such that you do not need a profound knowledge of machine learning techniques to get the most out of ML. Sure you have advanced options available on the service, but in our case, you will not need them. It creates predictive models quickly thanks to its dominant "1 Click" feature. BigML also offers a service which is cloud-based and highly scalable, that is easy to use, seamless to integrate, and instantly actionable. It works with less and more data so that everyone can implement decision-making data-driven in their applications.

Type: Web Analytics

Released on August 2014.

GUI Supported: Yes

Features:

- Support of WhizzML programming language
- Alexa voice service integration
- Bindings & libraries integration
- Open-source command line feature
- Google sheets add-on
- Native mac app support
- Predict server support
- Gallery support

Comparison:

- You can save the model on cloud and use it from any device you want as compared to BigML analytics software, which don't provide this feature.
- The UI/UX of the website is easy to use.
- It supports multiple languages as compared to BigML, which don't support.
- Provides almost all Machine Learning algorithms, which are also optimized.

Access/Source: Closed Access/Open Source.

License: Paid/Free.

API Supported: https://bigml.com/api

Platform: Windows, Mac, Web Based.

Community: https://blog.bigml.com/tag/community/

Videos: https://www.youtube.com/watch?v=CRxreWDFHVg

Reviews: https://www.g2crowd.com/products/bigml/reviews

Website: https://bigml.com/

Paxata

Paxata is one of the software, which focuses on only on data cleaning, and preparation, and this organization do not focus on the machine learning or statistical modelling part. It is a ms-excel like application that is easy to use, and with visual guidance making it easy to bring the data at one place, find and fix duplicate or missing data, and it also share and re-use data projects across teams. Paxata eliminates coding or scripting, to overcoming technical barriers involved in handling data. It uses a wide range of sources to acquire data and perform data exploration using powerful visuals which allowing the user to identify gaps in data quickly.

Type: Web Analytics

Founded by Chris Maddox, Dave Brewster, Nenshad Bardoliwalla, Prakash Nanduri In May 29, 2012.

GUI Supported: Yes

Features:

- Data exploration and profiling
- Data curation and governance
- Data source access and connectivity
- Data transformation, blending and modeling
- Automated Data Cleaning and Processing

Comparison:

- Visualize distributions in large data sets effectively, which enable the user to quickly spot outliers and treat them appropriately as compared to other analytical software.
- Provides recommendation to merge datasets based on matching column values as compared to other software.
- The cluster and edit feature in my opinion is its most powerful feature and reduces cardinality in column with text.

Access/Source: Closed Access/Closed Source.

License: Paid

API Supported: Yes

Platform: Web Based, Windows, Mac

Community: https://paxata.desk.com/

Videos: https://www.youtube.com/watch?v=FlMRl0p6w7k

Reviews: https://www.trustradius.com/products/paxata/reviews

Website: https://www.paxata.com/

Predixion

Predixion provided a platform for predictive analytics and established on the idea that it has the potential to create an intelligent, safer and better world and access to that power should not limit to a picked few with extensive statistical information. To accomplish that concept, Predixion formed a self-service predictive analytics program called 'Predixion Insight' that integrated with the entire predictive method. It is designed for market analysts and other non-technical users to allow the broader selection of predictive analytics. In addition, Predixion expedites the "Last Mile of Analytics" the deployment of robust predictions straight to the people who need them to take effect so the value of being predictive realized instantly. There is so many corporations rely on Predixion to help drive better decisions every day.

Type: Web Analytics

Founded by Jamie MacLennan, Simon Arkell, and Steve DeSantis in 2009.

GUI Supported: Yes

Features:

- API support for easily work on the data.
- Activity dashboard support
- Ad hoc reporting
- Provide business analysis
- Business intelligence support
- Charting the data support
- Provide customizable reporting

Comparison:

- Real Time and Historical data blending feature support as compared to Paxata.
- It allows the user created rules or custom rules.
- Real time advance analytics models at the edge feature as compared to other analytical software, which don't support this feature.
- Flexibility, scalability and user friendly.
- Marketing optimization, fraud detection, and guidance wizards as compared to paxata which don't support this feature.

Access/Source: Closed Access/Closed Source

License: Paid

API Supported: Yes

Platform: Web Based, Windows, Linux, Mac

Community: https://www.facebook.com/PredixionSW/

Videos: https://www.youtube.com/watch?v=3v0PqsQXt2w

Reviews: https://www.predictiveanalyticstoday.com/predixion-insight/

Website: https://greenwavesystems.com/

H2O

It is an open-source software, which is specialized for big-data analysis, H2O can be a strong influencer. Launched by H2O.ai, it allows thousands of potential models to derive different data patterns. It developed by Arno Candel in Java, Python, and R, the H2O software can be run from different environments like R and Python and has an easy-to-use WebUI, with the ability to visually inspect models during training. It is used in cloud computing systems-based datasets for exploring and analyzing them, in the Hadoop Distributed File System as well as in the conventional operating systems.

Type: Platform

Founded by Arno Candel in March 2015.

GUI Supported: No

Access: Open-sourced

License: Apache 2.0

Language: Java, Python, R

API Supported: Yes

Platform: Linux, Windows, mac

Community: https://www.h2o.ai/community/

Video: https://www.youtube.com/watch?v=9W_c2Ec23PM

Review:https://www.infoworld.com/article/3236048/machine-learning/review-h2oai-automates-machine-learning.html

Website: http://www.h2o.ai/

Repository: https://github.com/h2oai/h2o-3

PredictionIO

Built on top of a state-of-the-art technology, Apache PredictionIO is an open source machine-learning server for data scientists and developers to create predictive engines for machine learning tasks. It implements lamba architecture, which lets you evaluate and deploy search engines. PredictionIO can be integrated with other machine learning frameworks to accelerate learning process for managing infrastructure. Its distinguished features are prediction performance and the curse of dimensionality which allows the user to easy work on the prediction task.

Type: Library

Released on 2014

GUI Supported: No

Access: Open-sourced

License: Apache 2.0

Language: Java, PHP, Ruby, Python

API Supported: Yes

Platform: Linux, Windows, mac

Community: https://groups.google.com/forum/#!forum/theano-users

Video: https://www.youtube.com/watch?v=OhqBECqifhA

Review: https://www.g2crowd.com/products/predictionio/reviews

Website: https://predictionio.apache.org/

Repository: https://github.com/apache/predictionio

KNIME

Konstanz Information Miner or KNIME is a powerful data analytics platform. It is open-sourced, written in Java and released under GNU General Public License. It has multiple extensions that is KNIME server lite, KNIME team-space, KNIME server and KNIME web portal. The tool uses data pipelining concept for integration of various modules and offers visual interaction to the users.

While KNIME has been extensively used in pharmaceutical research, it is also popular among the software vendors, banks, firms and other research fields. Major functionalities are support for most of the file formats and databases, dynamic transformation and data blending, scoring through credit and access to over 1000 modules.

Type: Software

Foundation: 2011

GUI Supported: Yes

Access: Open-sourced

License: GNU General Public License

Language: Java

API Supported: Yes

Platform: Linux, Windows, OS X

Community: https://www.knime.com/knime-community

Video: https://www.youtube.com/watch?v=ft7Ksgss3Tc

Review:https://www.gartner.com/reviews/market/data-science-machine-learning-platforms/vendor/knime/product/knime-analytics-platform

Website: http://www.knime.org/

Repository: https://github.com/knime/knime-core

RapidMiner

Another popular integrated suite is RapidMiner, providing an environment for preparing data, developing and accessing models. While being designed on an open core model, it has both free and commercial releases. The free edition consists of deployment on a single logical server and is available under AGPL license. The proprietary edition is offered according to the size of enterprise.

Intelligence Unit of TU Dortmund initially released rapid miner as YALE in 2001. The rapid miner was written in Java, and it has a client/server model. By delivering an easy-to-use GUI and template-based frameworks, it minimizes the need to write code. In a survey earlier by Rexer Anslytics, RapidMiner received one of the highest satisfaction ratings.

Type: Software

Foundation: 2006

GUI Supported: Yes

Access: Free and Paid

License: AGPL, Proprietary

API Supported: Yes

Platform: Linux, Windows, mac

Community: https://community.rapidminer.com/

Video: https://www.youtube.com/watch?v=5tMAUokQK14

Review: https://www.g2crowd.com/products/rapidminer-studio/reviews

Website: http://rapidminer.com/

Repository: https://github.com/rapidminer/

Tableau

Tableau is mostly used for data analysis and visualization. Its aim is to help people understand the data better. It is much faster than the alternative solutions offering visual interaction. The license is proprietary and it comes in three editions professional, personal and public, with public edition restricted from saving visualizations locally. It has a rich community where stories with elegant data representations are shared. Connection to cloud data is also provided with ability to perform analysis of massive data in seconds.

Type: Software

Released in 2003

GUI Supported: Yes

Access: Free and Paid

License: Proprietary Software

API Supported: Yes

Platform: Linux, Windows, mac

Community: https://www.tableau.com/community

Video: https://www.youtube.com/watch?v=jj6-0cvcNEA

Review: https://www.pcmag.com/article2/0,2817,2491943,00.asp

Website: http://www.tableau.com/

Repository: https://github.com/tableau

Angoss

Angoss Predictive Analytics is a commercial software globally known for delivering valuable insights to large and medium scale businesses. Angsos' users are spread throughout different fields from finance to technology. It mainly offers five distinct products knowledge reader, knowledge seeker, knowledge studio, knowledge excelerator, and strategy builder with an individual subscription model for each.

While knowledge reader and knowledge excelerator both focus on visual data discovery, knowledge seeker and strategy builder are targeted towards data mining. Knowledge studio offers an integration of data mining and predictive analytics with several special features such as score validation, predictive modelling and scoreboard development.

Type: Software

Released in 1984

GUI Supported: Yes

Access: Paid

License: Proprietary Software

API Supported: Yes

Platform: Linux, Windows, mac

Community: http://www.angoss.com/community/knowledgeshare/

Video: https://www.youtube.com/watch?v=14JQHKa8ahk

Review:https://www.gartner.com/reviews/market/data-science-machine-learning-platforms/vendor/angoss/product/angoss-knowledgestudio

Website: http://www.angoss.com/

Splunk

Splunk is a commercial software with real-time analytics available through a searchable repository. It is designed to connect log and event management to find and debug the problems. With subscription-based model being the predominant source of revenue, splunk also avails a free trial with limited features.

Serving more than 10,000 customers worldwide, splunk has a library of more than 200 extensions to customize the type of analysis or inquire about a particular type of log information. With features such as real-time monitoring and automatic detection of outliers, it is specifically aimed for large enterprises.

Type: Software

Released in 2003

GUI Supported: Yes

Access: Paid

License: Proprietary Software

API Supported: Yes

Platform: Linux, Windows, mac

Community: https://www.splunk.com/en_us/community.html

Video: https://www.youtube.com/watch?v=rvjW5LJ0vbU

Review: https://www.softwareadvice.com/bi/splunk-enterprise-profile/

Website: http://www.splunk.com/

Repository: https://github.com/splunk

SAS Predictive Analytics

SAS offers a Software suite commercially available for all types of enterprises and personal use. It is the largest analytics vendor, independently operating in the business intelligence field. SAS business intelligence broadens the reach to insightful data with the help of guided analytics. It also roots out the production failures and factors affecting the quality of insights, with an emphasis on understanding the behavior of a consumer.

SAS software suite includes SAS visual analytics, SAS enterprise miner, SAS model manager, SAS high performance data mining, SAS scoring accelerator and SAS decision management. A visual interface is available with advanced options such as SAS language for developers. SAS also has macros to improve code efficiency and perform repetitive tasks easily. The only limitations are high cost and cross-platform compatibility.

Type: Software

Released in 1976

GUI Supported: Yes

Access: Free and Paid

License: Proprietary Software

Language: C

API Supported: Yes

Platform: Unix/Linux, Windows, IBM Mainframe, OpenVMS Aplha

Community: https://communities.sas.com/

Video: https://www.youtube.com/watch?v=K5qTKCzI2Ek

Review:https://www.g2crowd.com/products/sas-business-intelligence/reviews

Website http://www.sas.com/

Repository: https://github.com/sassoftware

IBM Watson

IBM Watson is a natural language-based question answering system. It uses automation of time bearing tasks such as predictive analysis and data preparation to quickly meet the user's need. Iterative learning is also used by obtaining insights from user interaction. The main focus is on understanding the question in depth to provide the answer. As it runs on cloud, accessibility through both on desktop and mobile is easy. The basic version of the software is free for all businesses, but the advanced features like extra storage and IBM analytics exchange data are paid.

Watson uses IBM's DeepQA and Apache unstructured information management architecture for operations, and apache hadoop to offer distributed computing. power servers avail parallel computing, with a vast collection of data from encyclopedias to news articles to build the knowledge. Watson was also used to challenge the successful contestants of a question answering game called Jeopardy, where it beat the contestants by over $50,000.

Type: Software

Released in 2010

GUI Supported: Yes

Access: Paid

License: Proprietary Software

Language: Java, C++, Prolog

API Supported: Yes

Platform: Linux

Community: https://www.ibm.com/communities/analytics/watson-analytics/

Video: https://www.youtube.com/watch?v=_Xcmh1LQB9I

Review: https://www.pcmag.com/review/352021/ibm-watson-analytics

Website: https://www.ibm.com/watson/

Repository: https://github.com/IBM-Watson

LIONSolver

If you need a predictive analytics software with data mining capabilities, LIONSolver is a possible solution. Although it is available with a proprietary license, a free version called LIONos. It has also been developed for academic use and non-profit research. LIONSolver assists in the decision making process based on large sets of data and deriving quantitative models from them.

The software utilizes self-tuning schemes while running different Machine Learning operations to learn from previous feedback and builds. Aside from allowing post-problem extensions, the latest version is compatible with most of the popular operating systems. The free version is also used to provide health care solutions.

Type: Software

Released on 2011

GUI Supported: Yes

Access: Free and Paid

License: Proprietary Software

API Supported: Yes

Platform: Linux, Windows, Mac OS X

Community:https://intelligent-optimization.org/LIONbook/lioncommunity.html

Video: https://www.youtube.com/watch?v=jj6-0cvcNEA

Website: http://lionoso.com/

Statistica

Statistica is a proprietary analytics software released by Dell. It incorporates data mining, machine learning, statistical and data analysis modules along with drag and drop components to create custom workflows. Initially created by statsoft, statistica was acquired by Dell, which later entered into an agreement with TIBCO to transfer the ownership.

Statistica has numerous products for specific applications, such as satistica automated neural networks, statistica data miner and statistica data warehouse. There are even separate editions according to the business size. Statistica library contains over 13,000 functions for different machine learning tasks. Both desktop and web versions are available to the users.

Type: Software

Released in 1991

GUI Supported: Yes

Access: Paid

License: Proprietary Software

API Supported: Yes

Platform: Linux, Windows, mac

Community: https://community.tibco.com/products/tibco-statistica

Video: https://www.youtube.com/watch?v=TtTfu30UmAg

Review: https://www.trustradius.com/products/tibco-statistica/reviews

Website: https://www.tibco.com/products/tibco-statistica

Repository: https://github.com/Statistica

Sisense

Sisense is a business Intelligence enterprise powered by single stack and in-chip technologies. It provisions a proprietary software with in-chip technology, crafted to maximize the CPU performance. The data preparation cost is minimized by its use even for vast or disparate data sets.

The company's in-chip technology integrates smart algorithms with the database to use in the in-chip cache, disk or ram, since the algorithms make intelligent resource allocation for memory. Sisense launched the analytics software Prism 10x specifically targeting the analysis of machine learning data at faster speeds. Later, Crowd Accelerated BI and Sisense 5 were released, the former to provide scalability for data, and the latter was released as an improvement to Prism 10x.

Type: Software

Released in 2004

GUI Supported: Yes

Access: Paid

License: Proprietary Software

API Supported: Yes

Platform: Linux, Windows, mac

Community: https://support.sisense.com/hc/en-us

Video: https://www.youtube.com/watch?v=mcnCEWQMwBY

Review: https://www.softwareadvice.com/bi/prism-profile/

Website: http://www.sisense.com/

icCube

icCube is a Business intelligence provider with proprietary solutions - Embedded and hosted business intelligence. There is support for multiple languages such as Java and R, and interactive dashboards that can be created and customized. The software can be installed locally or accessed from the cloud. icCube also provides consultation regarding big data-based decisions.

The BI solution is implemented in Java and runs inside an OLAP server, with a capability to use any kind of tabular data for analytics. Including the extensions such as Java and R interfaces, and function declarations, multidimensional expressions is employed as the query language.

Type: Software

Released on June, 2010

GUI Supported: Yes

Access: Free and Paid

License: Proprietary Software

API Supported: Yes

Platform: Linux, Windows, mac

Community: https://www.iccube.com/support/overview/

Video: https://www.youtube.com/watch?v=PJ8ofbYSbQE

Review:https://www.softwareadvice.com/bi/iccube-analysis-reporting-profile/

Website: https://www.iccube.com

Domo

Domo offers a cloud-based management platform with flexibility in deployment options, insightful analysis and an open data platform. It is also scalable according to the business demands. Domo was initially a London based company called Corda technologies, until entrepreneur Josh James bought and renamed it. Since then, it has received over $600 million of investment.

Domo is based on SaaS, with the ability to directly access data. For an enterprise, the insights can come directly from the interactions of the employees with Domo Buzz. Aside from being a mobile-first system, the support is exceptional. A separate database called Domo University exists just for the tutorials and learning the work process. Over a thousand apps are available on Domo Apps store, designed by third party developers.

Type: Software

Released on 2010

GUI Supported: Yes

Access: Paid

License: Proprietary Software

API Supported: Yes

Platform: Linux, Windows, mac

Community: https://dojo.domo.com/

Video: https://www.youtube.com/watch?v=DSV_VfUhOzs

Review: https://www.pcmag.com/article2/0,2817,2491955,00.asp

Website: https://www.domo.com/

Microsoft Excel

Excel is getting widely popular for its ability to extrapolate valuable insights and summarize data in creative ways. The inbuilt pivot tables and ease of use are also primary reasons for its popularity. The recently released version even offers recommendations for visualizing data based on the tabular operations performed by the user.

Using excel gives you an advantage of universal reach, extensions are available for all the major data analytics tools and seamless integration is possible for any device with Microsoft Cloud. The add-in power query helps with data discovery and mapping from text and XML files, Hadoop HDFS and web pages. Excel is packaged with Microsoft Office on windows, for which the proprietary license is required. Raw tabular data from excel can be easily imported into different visualization and analytical tools. Although the design is less intuitive, excel has a vast support community and plenty of tutorials available to learn from.

Type: Software

Released on 1987

GUI Supported: Yes

Access: Paid

License: Proprietary Software

API Supported: Yes

Platform: Windows

Community: https://techcommunity.microsoft.com/t5/Excel/ct-p/Excel_Cat

Video: https://www.g2crowd.com/products/microsoft-excel/reviews

Review: https://www.g2crowd.com/products/microsoft-excel/reviews

Website: http://office.microsoft.com/en-us/excel

Clear Analytics

An analytics system with powerful automation can rigorously accelerate generation of insights by eliminating the need for data preparation. Clear Analytics provides such feature. Their accuracy rate is claimed to be 100%, and with the visually interactive dashboard, even a beginner can generate advanced reports. It also incorporates a Self-service BI to answer our own questions. Consequently, there is better recovery and auditing and consolidation of data from anywhere.

The vision here is to connect the client and company directly by empowering the enterprises with easy to use sophisticated tools. With the extension of Microsoft Power BI, excel spreadsheets have advanced features like PowerPivot and Power Map. With data centralization, all the inputs are managed from a central location, so that the updating can be done in real-time. The pricing currently ranges between $100 to $499 for the license/subscription.

Type: Software

Released on 2017

GUI Supported: Yes

Access: Paid

License: Proprietary Software

API Supported: Yes

Platform: Linux, Windows, mac

Video: https://www.youtube.com/watch?v=U5gxKX8e1KI

Review: https://www.softwareadvice.com/bi/clear-analytics-profile/

Website: http://www.clearanalyticsbi.com/

Good Data

GoodData is a business intelligence company that offers platform as a service specializing in embedded analytics and intelligence. Founded by Roman Stanek in 2007, the company has surfaced as a leader, providing software to over 30,000 companies. The primary features are automated distribution and maintenance of multiple deployments over the cloud, responsive dashboards and visualisations with an option to collaborate with other users, email alerts and operational reporting. The business model is subscription based and available for all size of businesses.

Type: Software

Founded by Roman Stanek in 2007

GUI Supported: Yes

Access: Free and Paid

License: Proprietary Software

API Supported: Yes

Platform: Linux, Windows, mac

Community: https://support.gooddata.com/hc/en-us

Video: https://www.youtube.com/watch?v=fgkP0QMxq1w

Review: https://www.g2crowd.com/products/gooddata/reviews

Website: https://www.gooddata.com/

Repository: https://github.com/gooddata

Analance

Ducen's proprietary solution Analance is a business analytics software with advanced analysis capabilities. The distinctive feature of the software is its ability to tap users into multiple data sources, even IoT devices, so that insights that are more actionable can be generated from the various interactions.

Analance is designed in such a way that even a novice can operate the analysis flow and obtain valuable insights. Geo-Spatial and social analytics give an extra edge to the operable insights delivered by the software. Both desktop and web support are available, while only English is supported currently as the language. Pricing is quote-based and availability is limited to medium and large enterprises.

Type: Software

Released on 2003

GUI Supported Yes:

Access: Paid

License: Proprietary Software

API Supported: No

Platform: Linux, Windows, mac, Android

Video: https://www.youtube.com/watch?v=kHeYzU8Z6A0

Review: https://www.capterra.com/p/161816/Analance/

Website: https://analance.ducenit.com/

IBM Cognos

IBM Cognos is a web-based intelligence suite with self-service capabilities that allow the users to interact with insights even when they're offline. A toolset for monitoring significant events and generating reports is incorporated. There are several web elements cognos connection acts as the web portal, analysis studio is used for searching event related data or to perform single or multi-dimensional analysis on large data sets. Query studio to generate customized reports based on simple questions queried, report studio to produce comprehensive management reports with dynamic data, workspace to remotely collaborate and generate insights and event studio to be notified about the real-time changes in events.

Cognos also has components operated from the desktop, like IBM Cognos framework designer, map manager and transformer. The simple, intuitive interface, offline availability of the content, data protection and templates for generating reports make it an ideal solution for all types of businesses. Additionally, seven languages are supported while being compatible to all the popular operating systems.

Type: Software

Released on 2003

GUI Supported: Yes

Access: Free and Paid

License: Proprietary Software

API Supported: Yes

Platform: Linux, Windows, Solaris, AIX, HP-UX

Community: https://www.ibm.com/communities/analytics/cognos-analytics/

Video: https://www.youtube.com/watch?v=3NjOtSVwnG0

Review: https://www.softwareadvice.com/bi/ibm-bi-profile/

Website: https://www.ibm.com/products/cognos-analytics

Looker

Looker is a data discovery app entirely based on SQL. It is modeled using a language called LookML, which enables users to perform SQL operations without knowing much about it. Released in 2012 by Lloyd Tabb, the platform now has over $50 million funding. The software provides monitoring, sharing and reporting features to maintain the data consistency. Data modelling has also been developed to transform raw data into a much more understandable form with support for analysing big data.

Looker has a high user satisfaction rating with multiple positive social mentions. The enterprise plans for the software are quotable. Deployment is both cloud and open API based, besides multiple language support and cross platform compatibility. A self-service database is also available to ensure access to the support team.

Type: Software

Founded by Lloyd Tabb in 2012

GUI Supported: Yes

Access: Paid

License: Proprietary Software

API Supported: Yes

Platform: Linux, Windows, mac

Community: https://discourse.looker.com/

Video: https://www.youtube.com/watch?v=HBgJWCBOOZg

Review: https://www.g2crowd.com/products/looker/reviews

Website https://looker.com/

Repository: https://github.com/looker

Insight Squared

InsightSquared is a reporting and analytics suite for deep analysis into the business sales. Its pricing is divided into three plans basic, standard and enterprise, each costing $65 per user, $75 per user and $95 per user, primary differences being the team size limit and use of multiple data sources with custom access rules.

The software provided is scalable and has an option for customized configuration. multiple data connectors are also available for collaboration of data with other popular applications. The report builder delivers intuitive insights with the KPIs needed to boost the revenue of a business. Multiple languages are supported and mobile applications are also available for both Android and iOS.

Type: Software

Released in November, 2010

GUI Supported: Yes

Access: Free and Paid

License: Proprietary Software

API Supported: Yes

Platform: Linux, Windows, mac

Community:http://www.insightsquared.com/2014/08/business2-community/

Video: https://www.youtube.com/watch?v=SflEyX4khz8

Review: https://www.trustradius.com/products/insightsquared/reviews

Website: http://http//www.insightsquared.com/

JaperSoft

Owned by TIBCO, JaperSoft is a data integration platform with embedded business intelligence, analytics and reporting features. The first version was released in 2005 by Al Campa under the JaperSoft copyleft license, but later LGPL was applied. Adaptability to any business size, multi-dimensional analytics, partnership with SQL, dashboards with customization and the in-memory analysis engine make it popular among businesses. The pricing model is quote based.

Type: Software

Founded by Al Campa in 2003

GUI Supported: Yes

Access: Free and Paid

License: LGPL

API Supported: Yes

Platform: Linux, Windows, mac

Community: https://community.jaspersoft.com/

Video: https://www.youtube.com/watch?v=yRLvJgz9Dxk

Review: https://www.trustradius.com/products/jaspersoft/reviews

Website: https://www.jaspersoft.com/

Microsoft Power BI

Power BI is a cloud based integrated suite of analytics tools, which also incorporates self-service capabilities. The analysis can be done in real-time over the Power BI app or a browser without affecting the performance. Interactive dashboards with customization abilities are offered to the users along with built-in connectors for GitHub, Salesforce, Zen desk, etc.

With Power BI preview, events can generate reports with real-time insights. The refresh here is incremental, with a higher dataset refresh rate. After subscription to the proprietary license, there are two plans - Basic Plan with 1 GB data limit and the 10 GB per-user plan, with isolation from the noisy neighbors in both. Detailed documentation and video tutorials are available for users to understand the analytical workings.

Type: Software

Released on 2013

GUI Supported: Yes

Access: Paid

License: Proprietary Software

API Supported: Yes

Platform: Windows

Community: http://community.powerbi.com/

Video: https://www.youtube.com/watch?v=gqO0EiCn4cY

Review: https://www.pcmag.com/article2/0,2817,2494375,00.asp

Website: https://powerbi.microsoft.com/en-us/

Repository: https://github.com/Microsoft/PowerBI-visuals

MicroStrategy

MicroStrategy is a low-cost business intelligence software with integrated analytical capabilities and enterprise-level security. With micro strategy server and cloud, many analytics services can be deployed on the premises, which helps in as semblance of the project metadata. The data visualizations are interactive and flexible while the newly integrated micro strategy desktop and micro strategy web make it seamlessly easy for users to obtain insights from the data. Several featured mobile apps are also available to ensure the high-performance analytics along with real-time report generation. The subscription is proprietary with a quote-based system.

Type: Software

Released in 1989

GUI Supported: Yes

Access: Free and Paid

License Proprietary Software

API Supported: Yes

Platform: Linux, Windows, mac

Community: https://community.microstrategy.com/s/

Video: https://www.youtube.com/watch?v=MF4-7oZYab0

Review: https://www.g2crowd.com/products/microstrategy/reviews

Website: http://www.microstrategy.com/

MITS

MITS distributor analytics offers analytical solutions for enterprise resource planning and other similar businesses. With cloud-based report generation, bulk distribution owners can identify opportunities to increase revenue and efficiency. With over 200 customizable templates, most of the distributor queries related to purchasing, inventory and sales can be answered. It aims to minimize the guesswork by tapping into the insights strategically and meshing with the ERP System. While the license is proprietary, there is a free trial available with limited functionalities.

Type: Software

Founded in 2009

GUI Supported: Yes

Access: Paid

License: Proprietary Software

API Supported: Yes

Platform: Linux, Windows, mac

Community: https://help.mits.com/hc/en-us

Video: https://www.youtube.com/watch?v=AwZpz5lyOsw

Review: https://www.softwareadvice.com/bi/mits-profile/

Website: https://mits.com/

Oracle BI

The Oracle Business Intelligence Suite has Oracle's individual analytics applications into a single edition with reporting, planning, real-time analysis, in-memory analysis, performance tracking, report publishing, and mobile features. Instant analytics are available on the centralized dashboards, with collaboration and self-service capabilities along with protection from the oracle mobile security suite.

Various pre-built functions are available to ease the workflow for data analysts and get the answers to the prominent statistical questions. With the blending of communal and local data, meaningful insights can be generated in no time. The pre-processor license costs $300,000, while the named user plus license is around $4,500 per year, including the customer support.

Type: Software

Released on 2000

GUI Supported: Yes

Access: Paid

License: Proprietary Software

API Supported: Yes

Platform: Linux, Windows, Solaris, AIX, HP-UX

Community:https://community.oracle.com/community/business_intelligence

Video: https://www.youtube.com/watch?v=hlMtNVWW-7M

Review:https://www.trustradius.com/products/oracle-business-analytics/reviews

Website:https://www.oracle.com/in/solutions/business-analytics/business-intelligence/index.html

Pentaho

Pentaho is a BI endeavor acquired by Hitachi data systems. With the two editions - community and enterprise the functionalities such as creation of information dashboards, data reporting and data mining are offered. The enterprise edition has an annual subscription that is quote based.

Pentaho offers SQL Scripting as a perk with which creating pivots and objects while linking to different reports becomes seamlessly easy. Scheduling tasks with bursting reports by email can also prove highly productive for the user. Pre calculated and tabulated functions, complex aggregations, and an additional layer of security with Eclipse are some other primary advantages. Major cons would be limitation of the degree of automation and the relative difficulty in learning the Mondrian code.

Type: Software

Released in 2004

GUI Supported: Yes

Access: Free and Paid

License: Apache 2.0, Proprietary Software

API Supported: Yes

Platform: Linux, Windows, Mac OS X

Community:https://community.hitachivantara.com/community/products-and-solutions/pentaho

Video: https://www.youtube.com/watch?v=J8NbYQaQiPo

Review: https://www.trustradius.com/products/pentaho/reviews

Website: http://www.pentaho.com/

Repository: https://github.com/pentaho

Qlik Sense

Qlik, initially known as Quik, an acronym for "quality, understanding, interaction and knowledge", is a major provider of business intelligence and data visualisation software QlikView and QlikSense. The QlikSense has a user-friendly interface with a capability to combine the data sources into a single view. Visualisations are drag and drop in QlikView, and the self-service ability helps to get data extracted in an organised manner. Qlik is highly suitable for organisations with huge datasets.

Other prominent features include multi-source data integration, storytelling with data, reliability with multiple connections and smart search ability. With the self-service ability, there is no need to build queries or wait for the experts to extract insights. The associative engines in QlikSense exhaust all possible relationships in the data, so that insights can be generated faster.

Type: Software

Released in 1993

GUI Supported: Yes

Access: Free and Paid

License: Proprietary Software

API Supported: Yes

Platform: Linux, Windows, mac

Community: https://community.qlik.com/community/qlik-sense

Video: https://www.youtube.com/watch?v=jj6-0cvcNEA

Review: https://www.softwareadvice.com/bi/qlik-sense-profile/

Website: https://www.qlik.com/us/products/qlik-sense

PredicSis.ai

PredicSis.ai is a platform for predictive analytics, which provides support to improve your sales and marketing productivity through actionable perspicacity. It brings you essential predictions on each of your clients. All this information can inject into your CRM or marketing tools, and it discovers the reasons for your customer's behavior by the real-time predictions performance. A machine-learning software uses supervised learning algorithms to create models. It can be managed to gather a significant amount of aggregated data from many different sources, and this broad set of information is played an essential role in making decisions.

Seldon

Seldon is a predictive platform that gives content recommendations that formed on a Kubernetes cluster. Kubernetes is an open-source platform created by Google to program and its deployment, scaling and monitoring of applications, which are packaged in containers, hosted in the cloud and must of computing. The Kubernetes packaging provided to allows users to take the forms and execute them to any platform, which can be Azure, Amazon Services, Cloud Platform, etc. Seldon will enable users to capture and record user actions through their REST API, and then use that information to deliver personalized recommendations to other users.

Sisense

Sisense is a business intelligence platform that provides advanced analytical tools, which are used to manage and support the business data with visuals reporting, and analytics. The solution allows businesses to analyse vast and disparate datasets and create relevant industry trends for them. It will enable the enterprises to combine data from many different sources and combine them into a single database. Once done, the solution itself rearranges data into a predefined standard format. Users can then perform slicing and dicing over the complete data set using multiple filters and built-in analytic tools. Sisense includes much functionality for their scorecards and dashboards such as a query and report writer, data warehousing, data extraction, transform and load the data.

QlikView

QlikView is a platform for business intelligence, and it is one of the most traditional tools in business intelligence industry around the world. Deriving business insights and presenting it excellently, that what this tool does. With its state of art visualization capabilities, you would be amazed by the amount of control you get while working on data. This platform has an inbuilt recommendation engine, which is for an update you from best visualization methods in time to time while working on data sets. It allows you to share your data to get some insights and write some codes even to get accurate calculations on available data and then visualize it in various ways. You can consider it as a reporting tool with more intelligence and capabilities.

Computer Vision

AForge.Net

AForge.NET is a platform provides an open source C# framework which is designed for researchers and developers in the areas of Computer Vision and Artificial Intelligence which are image processing, neural networks, genetic algorithms, robotics, fuzzy logic, machine learning, etc. The framework included by the set of libraries such as image processing library, computer vision library, the neural network library, and machine learning library. It is a .NET framework for developers, allow to creating neural networks by software creators, computer vision, and semi-autonomous statistical analysis.

Category: Computer Vision.

Type: Framework.

Founded by Andrew Kirillov in December 21, 2006.

GUI Supported: No

Features:

- Library with image processing routines and filters.
- Support of computer vision library.
- Different set of library for video processing.
- Neural network computation library
- Support of evolution programming library.
- Support of libraries providing some robotic kits.

Comparison:

- AForge has many filters and is probably excellent as compared to OpenCV.
- Combine AForge with a good linear algebra library.
- Its "Image Processing Lab" makes much sense for filtering options (edge detection, thresholds, and so forth) and easing viewing functionalities.
- AForge is easy to work and easy environment as compared to OpenCV.

Access/Source: Closed Access / Open-Source.

License: LGPL v3

Language: C#.

API Supported: No.

Total Users/Projects: http://www.aforgenet.com/framework/projects.html

Platform: Windows, Linux and Mac.

Community: http://www.aforgenet.com/forum/

Videos: https://www.youtube.com/watch?v=OjeCFUzUkVQ

Reviews:http://www.discoversdk.com/products/aforge.net-framework-2.2.5#/overview

Website: http://www.aforgenet.com/framework/

CV Sandbox

Computer Vision Sandbox is a platform which provides a software package, that aims to allow solving different computer vision tasks such as computer vision based automation or robotics, video surveillance, and various sorts of image/video processing, etc. The plug-ins for video source plays the primary role in the system. There are many types of plug-ins to access different kinds of cameras, like USB web cameras, video capture boards, surveillance cameras, etc. Once the video is received, and it performs different image and video processing with an excellent variety of plug-ins. Combined those allow achieving many exciting results.

Category: Computer Vision.

Type: Software Package.

Released on March 27, 2015.

GUI Supported: Yes.

Features:

- Camera agnostic
- Multiple camera views
- Video recording
- Time lapse image snapshots
- Embedded scripting for advanced processing
- Wide range of image processing plug-ins
- Color filtering
- Objects counting
- Detection of square binary glyphs
- Bar codes detection
- Paralleling video processing
- Provides virtual camera device

Comparison:

- API supported feature as compared to other software packages.
- Detection in many areas like barcode, square binary glyphs.

Access/Source: Closed Access/Open Source

License: Free

Language: C#.

API Supported: Yes (cam2web)

Total Users/Projects: https://github.com/cvsandbox

Platform: Windows.

Community: No

Videos: https://www.youtube.com/channel/UCyJtgpnCUgFasbAgLW4hpAw

Reviews: http://weblisting.freetemplatespot.com/cvsandbox.com

Website: http://www.cvsandbox.com/

GRATF

It stands for Glyph Recognition and Tracking Framework. It is an open-source computer vision library. GRATF project was created to give a library, which makes recognition, pose estimation and localization of optical glyphs in still photos and video files. This library can also use in robotics applications for example, where glyphs may serve as commands or directions to robots. The most popular form of optical glyph recognition is augmented reality.

Category: Computer Vision

Type: Library

Founded by Andrew Kirillov in November 5, 2010.

GUI Supported: Yes

Features:

- It supports localization, recognition and pose estimation of optical glyphs in still images and video files.
- Used in robotics applications.
- Supports Augmented Reality in 2D and 3D.

Comparison:

- It Supports Glyph Recognition as compared to other library.
- It supports 3D Pose Estimation as compared to other library.
- It supports 3D Augmented Reality.

Access/Source: Closed Access/Open Source.

License: GPL License

Language: C#

API Supported: No

Total Users/Projects: https://code.google.com/archive/p/gratf/

Platform: Windows.

Community: http://www.aforgenet.com/forum/

Videos: https://www.youtube.com/channel/UCUNeuRG7X6md4TeM1b1RvVg

Reviews: https://www.codeproject.com/Articles/258856/From-glyph-recognition-to-augmented-reality

Website: http://www.aforgenet.com/projects/gratf/

PCL

The PCL or Point Cloud Library is a computer vision framework, and it is a large-scale open source project for 2D/3D image and point cloud processing. The pcl framework, which contains various state-of-the-art algorithms, that includes filtering, surface reconstruction, feature estimation, segmentation, registration, and model fitting and many other to use in computer vision. For example- these algorithms are used to filter outliers from noisy data, stitch 3D point clouds together, extract critical points and based on their geometric appearance it compute descriptors to recognize the objects in the real world, and create surfaces from point clouds and visualize them.

Point Cloud Library is cross-platform and has been compiled and deployed successfully on all platform like Linux, Mac, Windows, and Android/iOS. The point cloud library is split into a series of smaller code libraries that can be compiled separately to simplify the code and efficiently work on it. Different way to think about Point Cloud Library is as a graph of code libraries, related to the Boost set of C++ libraries.

Type: Library

Founded by Willow Garage in March 2010

GUI Supported: No

Features:

- It supports many features like filters, features, key points, registration, and segmentation.
- 3D Point Cloud
- Mapping and Localization
- Aerial Point Cloud
- Face Scan
- Object Detection
- Shape Detection and Recognition
- Support 2D and 3D Images

Comparison:

- PCL library supports 3D Sensors as compared to Open CV, which does not support 3D.

- Support 3D Visualization as compared to other library.

Access/Source: Open Access/Open Source

License: BSD License

Language: Python, C#

API Supported: Yes

Platform: Windows, Linux, Mac, Android.

Community: http://pointclouds.org/news/2013/04/16/community-resources-and-repositories/

Videos: https://www.youtube.com/watch?v=GRchtJUgyPA

Reviews: No

Website: http://pointclouds.org/

Rvision

Rvision is a computer vision library for R, and it is small but growing library. The Rvision library is based on the extensive OpenCV library for C/C++, and it is the state-of-the-art method for computer vision in the open source world. The primary objective of Rvision is to give R users with all the primary functions to read and manage images, videos and real-time camera streams, with importance on speed and accuracy. The Rvision library is different from all the other image manipulations packages for R that all the other packages are limited in their processing speed or volume, and cannot immediately access the frames from pictures, videos or camera streams.

Type: Library

Founded by Simon Garnier in July 6, 2016.

GUI Supported: No

Features:

- It is an opencv library for R users, which supports computer vision functions.
- Supports real time camera streams.
- Read and manages the images and videos.

Comparison:

- Rvision is faster and accurate as compared to other R image manipulation packages.
- Other R packages are limited in their processing, volume and speed and cannot immediately access the frames.

Access/Source: Open Access/Open Source

License: GPL-3

Language: R

API Supported: No

Platform: Windows, Linux

Community: No

Videos: No

Reviews: No

Website: https://swarm-lab.github.io/Rvision/

VLFeat

The VLFeat is a free open-source computer vision library, which implements popular computer vision algorithms specializing in local features matching and extraction, and image understanding. There are many algorithms in VLFeat library for the work on computer vision applications some of them are fisher vector, k-means, sift, mser, vlad, hierarchical k-means, slic super pixels, agglomerative information bottleneck, quick shift super pixels, large-scale svm training, svm, and many others. The VLFeat library is written in C for efficiency and compatibility, with interfaces in matlab for ease of use, and detailed documentation throughout.

Type: Library

Founded by Andrea Vedaldi and Brian Fulkerson in 2007.

GUI Supported: No

Features:

- Local Features Matching and Extraction.
- The library comes with the support of Fisher Vector, K-means, SIFT, MSER, VLAD, hierarchical k-means, SLIC super pixel.
- This library is for Matlab users.

Comparison:

- This library supports SIFT for better object detection as compared to other libraries, which doesn't support this features.
- The processing speed and accuracy is high as compared to other computer vision libraries.

Access/Source: Closed Access/Open Source

License: BSD

Language: C++, Matlab

API Supported: No

Platform: Windows, Mac, Linux

Community: No

Videos: https://www.youtube.com/watch?v=bbPsz8mloTY

Reviews: https://www.g2crowd.com/products/vlfeat/pricing

Website: http://www.vlfeat.org/

BoofCV

BoofCV is an open source library for Java, which is used for real-time robotics, image processing and computer vision applications. This library is written from scratch for easy understanding and high performance. Its working includes a wide variety of subjects including, camera calibration, optimized low-level image processing, feature detection and feature tracking, recognition, and the making of structure from motion. This library has been released for both academic and commercial use under an Apache 2.0 license. BoofCV library is formed into several packages, which include geometric vision, image processing, features, calibration, recognition, and visualize.

Type: Library

Founded by Peter Abeles in November 8, 2011.

GUI Supported: No

Features:

- Image Processing (blur, edge, binary, enhancement, 360 photo).
- Segmentation (super pixels, thresholding, color).
- Detection (corner, SURF, SIFT, line, shapes).
- Image Association (nearest neighbor).
- Tracking (KLT, object tracking, motion detection).
- Camera Calibration (chessboard, circles, squares).
- Recognition (QR Code, CNN, fiducial markers).
- Structure from Motion (stereo disparity, mosaic/stabilization),

Comparison:

- Real-time computer vision and robotics applications as compared to other libraries, which do not support real time analysis
- This java library is comes with geometric vision, image processing, features, calibration, recognition, and visualize.

Access/Source: Closed Access/Open Source

License: BSD

Language: Java

API Supported: http://boofcv.org/javadoc/index.html?help-doc.html

Platform: Windows, Linux

Community: No

Videos: https://www.youtube.com/watch?v=9G51ZX1Eov8

Reviews: https://sourceforge.net/projects/boofcv/reviews/

Website: https://boofcv.org/index.php?title=Main_Page

ChainerCV

ChainerCV is a free open source deep learning library for computer vision, which supports numerous neural network models as well as software components needed to research in computer vision. These implementations emphasize integrity, versatility and good software engineering applications. The library is created to efficiently work on the deep learning application, and the tools in the library can be used as a baseline for future research in computer vision. ChainerCV supports algorithms to solve tasks in the computer vision field such as object detection while considering usability and predictable performance as the top priorities. This library makes it perfect to be used as a building block in more massive software projects such as robotic software systems even by developers who are not computer vision experts.

Type: Library

Released by Toru Nishikawa and Daisuke Okanohara in June 2015.

GUI Supported: No

Features:

- Image Classification (ResNet, VGG)
- Object Detection (Faster R-CNN, SSD, YOLO)
- Semantic Segmentation (SegNet, PSPNet)
- Instance Segmentation (FCIS,)

Comparison:

- Implementations of computer vision networks with a cohesive and simple interface as cv sandbox do not support it.
- Training scripts that are perfect for being used as reference implementations.
- Tools such as data loaders and evaluation scripts that have common API.

Access/Source: Open Access/Open Source

License: MIT License

Language: Python

API Supported: Yes

Platform: Windows, Linux

Community: https://github.com/chainer-community

Videos: https://www.youtube.com/watch?v=DGJazkyw0Ws

Reviews: No

Website: https://chainer.org/

EmguCV

Emgu CV is an open source cross-platform Open CV image-processing library for the .NET platform. This library is allowing OpenCV functions to be called from .NET compatible languages such as c#, vb, vc++, iron python etc. It can be compiled in mono and run on windows, linux, mac, iPhone, iPad and android devices. One of the EmguCV goals is to provide a simple to use computer vision infrastructure for .NET programmers that help them build reasonably sophisticated vision applications quickly. This library is used in many areas in vision, stereo vision and robotics, including factory product inspection, camera calibration, and medical imaging user interfaces. EmguCV also wraps a full general-purpose machine-learning library from the OpenCV image-processing library.

Type: Library

Released in February 27, 2012.

GUI Supported: No

Features:

- Image class with generic color and depth
- Automatic garbage collection
- XML serializable image
- XML documentation and intelligence support
- The choice to either use the Image class or direct invoke functions from OpenCV.
- Generic operations on image pixels

Comparison:

- EmguCV can run on Mono i.e. (*inx) now as compared to Aforge.
- Performance and Accuracy of EmguCV is better than Aforge.
- The amount of image processing functions is vast.

Access/Source: Closed Access/ Open or Closed Source.

License: GNU GPL license v3.

Language: .NET, C#, C++, VB.NET

API Supported: http://www.emgu.com/wiki/index.php/Documentation

Platform: Windows, Linux, Mac, Android, IPhone/IPad.

Community: http://www.emgu.com/forum/

Videos: https://www.youtube.com/watch?v=b8k7Bnb0nao

Reviews: https://sourceforge.net/projects/emgucv/reviews/

Website: http://www.emgu.com/wiki/index.php/Main_Page

OpenCV

A popular open source tool targeting Computer Vision is OpenCV. Initially developed by Intel as a research initiative in 2000, it is now freely released under the BSD license. It comprises of over 2000 algorithms, allowing a wide range of tasks to be accomplished. The library is applied to use in tech giants like IBM and Google and has a large community of users.

The goal of OpenCV is to advance the use of Computer Vision in commercial applications. It is highly efficient algorithmically for real-time programs. Several GPU interfaces like opencl have also been integrated recently to enhance the performance. It is compatibility with over 10 operating systems makes it an attractive option for developers.

Type: Software

Released in 2000.

GUI Supported: No

Access/Source: Closed Access/ Open or Closed Source.

License: GNU GPL license v3.

Language: C#, C++

API Supported: Yes

Platform: Windows, Linux, Mac, Android, IPhone/IPad.

Community: http://answers.opencv.org/questions/

Videos: https://www.youtube.com/watch?v=X6rPdRZzgjg

Reviews: https://www.g2crowd.com/products/opencv/reviews

Website: https://opencv.org/

Statistical Softwares

SPSS

SPSS Statistics is a leading statistical software, which is used to solve a kind of business, analytical and research problems. This software provides a type of techniques, which include hypothesis testing and reporting, and ad-hoc analysis, which makes it easier to manage data for future use like select and perform data for review, and share your results. New features in this software include Bayesian statistics, publication charts in the ready state and improved integration of third-party software. This software also offers a base edition, which provides for optional add-ons to expand predictive analytics capacities. This Statistics software proceeds to increase convenience to advanced analytics through advanced tools, output, and easy to use features. Now look at some new features you will see in this software that is designed to help you to create better predictive models. Estimate risk more accurately and works faster or improve analytical performance.

Type: Software

Founded by IBM Cooperation in 1968.

GUI Supported: Yes

Features:

- Multiple data sources support
- Visual analysis streams
- Automatic data preparation
- Automated modeling
- Algorithmic methods
- Text analytics
- Geospatial analytics
- Multiple deployment methods
- Combination of IBM predictive analytics software
- Multiple data sources & environments
- Flexible deployment options
- Open & integrated solution

Comparison:

- SPSS has an easy point-and-click interface, making data analysis relatively straightforward.
- Another benefit is the huge community as compared to other software, which don't have community.
- Support basic statistics, univariate and bivariate multivariate Statistics and special models.

Access/Source: Closed Access/Closed Source

License: Paid

API Supported: No

Platform: Windows, Linux, Mac

Community: https://developer.ibm.com/predictiveanalytics/2015/04/20/welcome-to-the-new-spss-community/

Videos: https://www.youtube.com/watch?v=eTHvlEzS7qQ

Reviews: https://www.g2crowd.com/products/ibm-spss-statistics/reviews

Website: https://www.ibm.com/analytics/spss-statistics-software

Minitab

Minitab is a statistical software, which provides features like data analysis and graphical data presentation. Minitab can perform a different kind of data analysis and presentation functions, including statistical analysis and graphical display of data. In this statistical software, you can instantly calculate regression, prepare different charts, and enter the data, which works very similar to MS-Excel. The new release of the Minitab includes advanced capabilities for statistical modelling and measurement systems analysis and Design of Experiment, and it delivers quality enhancement for the professionals or even higher insight into their processes, it also provides many features for enabling them to make better conclusions, faster and more accurately than ever before.

Type: Software

Founded by Barbara F. Ryan, Thomas A. Ryan, Jr., and Brian L. Joiner in 1972.

GUI Supported: Yes

Features:

- Analysis of variance (ANOVA)
- Statistical inference
- Correlation and regression
- Hypothesis tests and confidence intervals
- Measurement systems analysis
- Control charts
- Design of experiments
- Process capabilities

Comparison:

- Support of Industry-proven process improvement tools as other software doesn't support
- Support of Project Roadmaps and other tools as compared to other software.
- Provide Real-time Insights
- Provide Value Stream Mapping
- Provide Monte Carlo Simulation

- Provide Quality Function Deployment

Access/Source: Closed Access/Closed Source

License: Paid

API Supported: No

Platform: Windows

Community: http://www.minitab.com/en-us/support/

Videos: https://www.youtube.com/watch?v=q1zKe8u7NfM

Reviews: https://www.capterra.com/p/109731/Minitab-17/

Website: http://www.minitab.com/en-us/

Statistix

Statistix is a statistical software, which provides many features like statistics, analytical, and graphical analysis. Statistix software includes some additional features like import and export support for Excel and text files, powerful data manipulation tools, linear models such as linear/logistic/Poisson regression, nonlinear regression, time series, association tests, nonparametric tests, survival analysis, quality control, power analysis, and more. This software is integrated with all the basic and advanced or powerful data manipulation and statistics tools you need in a single, inexpensive package. This statistical software is easy to learn and simple to use. It saves your valuable time and money. This software comes with Strong transformations method for modifying and creating variables using algebraic expressions with over more than 50 built-in functions, constructs and supports flexible data types include real, integer, date, and string variables.

Type: Software

Released in 2013.

GUI Supported: Yes

Features:

- Performs sample test
- Performs hypothesis test
- Performs linear and nonlinear modals
- Performs analysis of variance
- Performs association test
- Performs normality test
- Performs time series
- Performs quality control

Comparison:

- Fast and comprehensive as compared to Minitab.
- Import/export support from Excel, text files and many more features also integrated as compared to Minitab, which don't provide such extra features.
- Combined basic and advanced statistical tools for predictions.

- Powerful data manipulation tools, which are easy to implement as compared to other softwares, which doesn't provide data manipulation tools.
- Easy to learn and use

Access/Source: Closed Access/Closed Source

License: Paid

API Supported: No

Platform: Windows

Community: No

Videos: https://www.youtube.com/watch?v=KOtIhVHozG8

Reviews: https://www.capterra.com/p/125836/Statistix/

Website: https://www.statistix.com/

Maxstat

Maxstat is a statistical software, which provides many features like statistics, analytical, and graphical analysis. Max stat software offers more than 100 statistical tests and tools which are commonly used in the study of the scientific data, and it includes descriptive or hypothesis, linear regression, non-linear regression, correlation, time series, multivariate analysis. This software even helps you to design your experiments by calculating the sample sizes of data and power. Max stat is an easy to use and an affordable statistics software. MaxStat software is a perfect software for students and young researchers because it guides users through each step of the analysis. This software can complete your statistical analysis within a single dialog box in three easy steps.

Type: Software

Released in Germany in 2009.

GUI Supported: Yes

Features:

- Support 100 statistical tests and 30 charting styles
- Reporting and testing hypothesis feature integration.
- Provide testing hypothesis and regression analysis
- Correlation analysis
- Multivariate analysis
- Time series
- Word processing and reporting

Comparison:

- Complete spreadsheet management and very intuitive with over 30 charting styles as compared to other softwares, which don't provide such features.
- Easy to use and short statistical analysis steps for the perfections.
- Easy to interprets and create high quality graphs, which has user friendly interface

Access/Source: Closed Access/Closed Source

License: Paid

API Supported: No

Platform: Windows

Community: No

Videos: https://www.youtube.com/watch?v=fxHQJhafY2Y

Reviews: https://www.capterra.com/p/143824/MaxStat/

Website: https://maxstat.de/en/home-en/

Acastat

Acastat is a statistical software, which provides many features like frequencies tables, correlations, non-parametric tests, cross-tabulations, t-tests, descriptive statistics, OLS and logistic regression, and more. It also works on format variable, value labels, and recode variables, use controls and set missing values. It is a statistical data analysis software, which has been designed for allowing statistical analysis quick and uncomplicated jobs. AcaStat for mac and windows will enable data to be extracted from spreadsheets, copy, pasted, dragged, and dropped to the tool.

This software analysis data performs a non-parametric test, t-test, correlation, regression, descriptive statistics etc. After data analyzed, and a summary is produced for the user to examine the results. The software also verifies calculations done manually. The software can also calculate the price elasticity of demand and queuing theories. For decision, making it can produce decision tables to be analyzed additional.

Type: Software

Founded by Acastat Company in 1999.

GUI Supported: Yes

Features:

- Perform statistical analysis
- Support of many features like statistical test, calculates significance and confidence intervals, compare data summaries etc.
- Different solution for iPad, windows and Mac
- IBook for easy guide is available free as compared to other softwares, which do not provide guides.

Comparison:

- Simple but complete software and low-cost as compared to ncss software.
- Drag and drop variable selection and with integrated charts modules.

Access/Source: Closed Access/Closed Source

License: Paid

API Supported: No

Platform: Windows, Mac

Community: No

Videos: No

Reviews: https://www.capterra.com/p/111619/AcaStat/

Website: http://www.acastat.com/

NCSS

NCSS is a statistical software, which has been committed to providing analysts, academics, researchers, scientists, and other experts with quality statistical features that is complete and perfect but still natural and easy to use. This software provides an entire collection of many sets of tools, which are used for graphical and statistical problems and easy-to-use to analyze and visualize data. This software is used by different organizations and individuals for the use their Data to import or enter their data to NCSS and use tools to analyze.

This software can also choose the columns and run the analysis to achieve accurate and easy to read numeric output such as graphics. It supports the imports of all essential types of data files quickly, or users can copy and paste their data, or enter it directly. One of the features of NCSS is it has excellent filtering and transformation features for managing your data. In this software, finding the right analysis or graphics procedure is easy using the drop-down menu. The report and graphical tools are easy to use and have built-in help messages for every option. All users need to do is to select the columns to be analyzed, choose the desired reports and plots, and click Run to obtain their results.

Type: Software

Founded by Jerry L. Hintze in 1981.

GUI Supported: Yes

Features:

- Easy to use to analyze and visualize data.
- Complete set of tools for statistical and graphical analysis
- Use in Medical investigation and business analytics.
- Support Quality Control.

Comparison:

- Complete and easy to use tools and easy analysis setup.
- Ready to use output and flexibility in reporting and graphing as compared to acastat which doesn't provide set of tools.
- Intuitive Data Management.

- Support different types of data files.

Access/Source: Closed Access/Closed Source

License: Paid

API Supported: No

Platform: Windows

Community: No

Videos: https://www.youtube.com/watch?v=MIaM0sehkrU

Reviews: https://www.predictiveanalyticstoday.com/ncss/

Website: https://www.ncss.com/

MATLAB

Matrix Laboratory or MATLAB is a proprietary numerical computing platform developed by MathWorks. It provides a desktop environment with tools for visualizing data and apps and add-ons for a wide range of machine learning problems. MATLAB is one of the most popular commercial platforms for data science due to the ease-of-use and its ability to scale and run on embedded devices.

Initially designed in late 1970s, MATLAB now has more than 1 million users from multiple technological domains. In addition to it having a very simple syntax based on an interactive mathematical shell, an interface to various programming languages like C, C++ and java is given. The pricing is dependent on the occupation and residency of the user.

Type: Software

Released in 1984

GUI Supported: Yes

Access: Paid

License: Proprietary Software

Language: C, C++, Java

API Supported: Yes

Platform: Linux, Windows, mac

Community: https://blogs.mathworks.com/community/

Video: https://www.youtube.com/watch?v=T_ekAD7U-wU

Review: https://www.capterra.com/p/125813/MATLAB/

Website: http://mathworks.com/products/matlab

GNU Octave

GNU Octave is a high-level programming language primarily designed for numerical computations. With this software, users can perform numerical experiments by matching the compatibility to MATLAB. It was initially developed in 1988 and is freely released under the GNU GPL license.

The primary benefits of using Octave are high compatibility with MATLAB scripts, detailed documentation, in-built visualization tools and portability. All the packages developed for it and are centrally maintained at Octave Forge. The main aim with which Octave was developed under the GNU project was to help users in solving realistic problems and in teaching and research applications.

Type: Software

Released in 1988

GUI Supported: Yes

Access: Free

License: GNU GPLv3

Language: C, C++, Java

API Supported: Yes

Platform: Linux, Windows, mac

Community: https://www.gnu.org/software/octave/support.html

Video: https://www.youtube.com/watch?v=sHGqwF2s-tM

Review: https://www.g2crowd.com/products/gnu-octave/reviews

Website: https://www.gnu.org/software/octave/

Wolfram Mathematica

Mathematica is a modern computational programming system provisioning a higher-level environment in which emphasis is on automation and user's ability to work on problems without knowing much about the intricate workings of the system.

Mathematica comes with sophisticated design and robust implementation of machine learning algorithms to handle large-scale computations. Results are provided in an intuitive manner with dynamic visualizations. It offers integration with cloud with ability to use real-time data. The pricing is subscription-based, depending on the personalized, cloud or enterprise edition.

Type: Software

Released in 1988

GUI Supported: Yes

Access: Paid

License: Proprietary Software

Language: Wolfram Language, C, C++, Java, Mathematica

API Supported: Yes

Platform: Linux, Windows, mac, Raspbian

Community: http://community.wolfram.com/

Video: https://www.youtube.com/watch?v=_P9HqHVPeik

Review: https://github.com/WolframResearch

Website: http://www.wolfram.com/mathematica/

Repository: https://github.com/WolframResearch

AI Platforms

SAP Net Weaver

Net Weaver is also known as the SAP OS layer, which is a combination of the SAP kernel, which is primarily the WEB AS and any SAP software tool for business enablement. Net Weaver is a technology of business suite, which is developed by the software firm SAP SE, and it is the scientific foundation for many SAP purposes. It is a resolution stack of SAP's technology products. It is the runtime conditions for the SAP plans, and all of the mySAP business suite runs on SAP WebAS like SRM, CRM, SCM, TMS, PLM, and ERP. It permits users to create new applications along with transferring data from existing systems. It is user-friendly because it can input both regular and verbal directions from users.

Type: Software

Founded by SAP on March 31, 2004.

GUI Supported: Yes

Features:

- People integration features are multi-channel access, portal, collaboration support.
- Information integration features are business intelligence, knowledge management, master data management support.
- Application Platform: ABAP Engine, Virtual Machine Controller, Java Engine, Business Services support.
- Composite Application Framework support.
- Database Administration support.
- SAP knowledge warehouse support.

Comparison:

- It provide multiple channel access as compared to openAI which don't provide multiple channel access.
- All of the mySAP business suite runs on SAP WebAS like SRM, CRM, SCM, TMS, PLM, and ERP.
- It can input both regular and verbal directions from users.

Access/Source: Closed Access/Closed Source.

License: Paid

Language: Java, .NET

API-Supported:
https://help.sap.com/viewer/8e0fc607142d4758895cc176c28cfdb5/7.1.18/en-US/13041975d8ce4d4287d5205816ea955a.html

Platform: Windows, Linux and Mac.

Community: https://www.sap.com/india/community/topic/netweaver.html

Videos: https://www.youtube.com/watch?v=S2caXvU_H7M

Reviews: https://www.g2crowd.com/products/netweaver/reviews

Website: https://www.sap.com/index.html

Tensor Flow

TensorFlow is an open-source software library used to perform numerical computation and dataflow programming across a range of tasks. Being a symbolic math library, it is also used for machine learning applications such as neural networks. It is heavily used at Google for both production and research, replacing its predecessor DistBelief.

Initially, the library was developed by the Google Brain team for internal Google use. But it was made open-source under Apache license on November 9, 2017. The flexible architecture and auto-differentiation allows users to construct the graphs and write the loops driving the computation. The computations are also deployable on single CPUs, server or mobile device with an API.

Tensorflow is applicable to range of domains, main use cases being voice and image recognition, text analysis, time series fluctuations analysis, video detection. It provides an API for multiple languages such as C++, Haskell, Java, Go, and Rust APIs. Third party packages are available for C#, Julia, R, and Scala.

OpenAI Universe

Universe is a platform specifically targeted on AI's general intelligence. Founded by the entrepreneurs Elon Musk and Sam Altman, OpenAI is a non-profit research-based company that is aiming to promote the development of a friendly AI that is also ethically balanced, in order to avoid scenarios such as intelligence explosion. All research and patents under OpenAI have been made publicly available.

OpenAI universe allows developers to train and asses AI agents in real-time and complex environments. Many executable classes are publicly available packaged inside Docker containers. A Universe environment mainly consists of a client and a remote, in which they communicate with each other using the VNC remote desktop system.

OpenAI Gym

OpenAI Gym is a collection of different environments, with an objective to standardize the definitions of any environment in research, so that reproducibility of publications can be increased. Written in Python, Gym is used for development and comparison of reinforcement learning techniques.

Gym is designed not to presume anything about the structure of AI agents. Currently, it is only available in Python, but will soon be available in other languages. One of its great benefits is the high compatibility factor with numerical computation libraries.

Google Cloud Machine Learning Engine

Machine Learning Engine provides the infrastructure to train and host models, with google prediction API features incorporated. Prediction APIs are a set of programming interfaces designed by google for developers to build, deploy and export machine learning models. Google provisions its state-of-the-art pattern matching and recognition capabilities with the APIs. They can be accessed through google cloud platform from the browser or mobile apps. Numerous machine learning models can be deployed easily on high-performance google clusters. The engine has tensorflow integration with pre-built training models.

Ayasdi

Another enterprise level machine-learning platform is Ayasdi. Known for its seamless integration and simple design, Ayasdi is equipped with a suite of smart and targeted applications to deliver the competitive advantage from a big data. It has the ability to simultaneously create, test and deploy machine-learning models at scalable level.

Founded in 2008, Ayasdi now incorporates more than 30 statistical and machine-learning algorithms unified with its mathematical framework Topological Data Analysis, producing a network similarity map as a result to minimize bias and provide helpful correlations. It has a vision of obtaining hidden insights from complex data easier for clients.

SAP Leonardo

SAP Leonardo is an Internet-of-Things portfolio incorporating management of big data and adaptive applications to build systematically next-generation products with enhanced connectivity. It was originally released in January, 2017 but was relaunched later as a "digital innovation system". While it has a moderate learning curve for using the environment, the option to create 3D projects and requirement of a lesser number of steps in any specific process makes it a popular choice for enterprises with an IoT environment.

Azure Machine Learning Studio

Azure is a prominent cloud computing service provided by Microsoft. Through Azure, software, platforms and infrastructure can be accessed as a service. Azure Machine Learning Studio is an interactive environment to design and experiment different predictive analysis models. For beginners, drag and drop datasets are available to be used in conjunction with analysis modules.

Azure Studio includes a comprehensive list of Machine Learning APIs and services. The collaborative environment, portability of code, compatibility with most of the operating systems, it is a well-favored choice among the machine learning enthusiasts.

Big Data Tools

MLBase

MLbase provides the power of machine learning for both frontend and backend users and ML researchers to the system harnessing. It provides a simple declarative way to specify ML tasks. A new optimizer to select and dynamically adapt the choice of the learning algorithm. A set of high-level operators to enable ML researchers to scalable implement a wide range of ML methods without in-depth systems knowledge and a new run-time optimized for the data-access patterns of these high-level operators. Many users do not understand the tradeoffs and challenges of choosing between different learning techniques and parameterizing by which the complexity of existing machine learning algorithms is often overwhelming.

Type: Library

Founded by the AMP Lab at UC Berkeley.

GUI Supported: No

Features:

- Data manipulation & preprocessing
- Score-based classification
- Performance evaluation (e.g. evaluating ROC)
- Cross validation and Model tuning (i.e. search best settings of parameters)

Comparison:

- It is developed for both front end and back end users as compared to sequencel, which doesn't provide both, ends supports.
- It is integrated with the Spark to work in big data and other big data related technologies.

Access/Source: Open Access/Open Source

License: MIT

Language: Spark

API Supported: No

Platform: Windows, Linux

Community: https://community.hortonworks.com/index.html

Videos: https://www.youtube.com/watch?v=IxDnF_X4M-8

Reviews: No

Website: http://mlbase.org/

SequenceL

SequenceL is a combination of compilers and toolset for auto parallelizing and functional programming. Its primary design objectives are the performance on a multi-core processor, platform portability or optimization, ease of programming, and language simplicity and readability. Its principal benefit is that it can be utilized to address straightforward system that automatically takes full use of all the processing power possible, without programmers requiring concerned and the challenges of standard directive-based programming approaches with identifying the parallelisms, specifying vectorization, bypassing race conditions, and other.

Type: Software

Founded by Dr. Daniel Cooke, Dr. Nelson Rushton, Texas Tech University, in 1989.

GUI Supported: No

Features:

- auto parallelizing and multi-core processor support
- portability or optimization and bypassing race conditions

Access/Source: Closed Access/Closed Source

License: Proprietary

Language: Python, Java, C++, C+ C#.

API Supported: No

Platform: Windows, Mac, Linux

Community: No

Videos: https://www.youtube.com/watch?v=DE8zBnzFlRQ

Reviews: No

Website: No

Spark MlLib

MLlib is a distributed machine-learning framework, which runs on top of spark because of the shared memory-based spark structure. According to the MLlib developers that the benchmarks, done against the alternating least squares implementations. Spark MLlib is nine times as fast as the apache mahout is (before mahout gained a spark interface). Spark is becoming the de-facto platform for developing machine learning algorithms and applications. Spark MLlib are implementing more machine algorithms in a scalable and compact manner by the developers to managing in the spark framework.

Type: Framework

Founded by Apache Software in October 15, 2012.

GUI Supported: No

Features:

- Shared memory-based and scalable and compact
- Contains machine-learning libraries and data frame-based API.

Comparison:

- Faster in processing as compared to mahout and faster than hadoop's map reduce.
- It learns from existing categorization and then assigns unclassified items to the best category.

Access/Source: Open Access/Open Source

License: Apache License v2.0

Language: Scala, Java, Python, R, Spark

Total Users/Projects: https://spark.apache.org/third-party-projects.html

API Supported: https://spark.apache.org/docs/2.2.0/mllib-guide.html

Platform: Windows, Linux, Mac

Community: https://spark.apache.org/community.html

Videos: https://www.youtube.com/watch?v=qKYpMPPL-fo

Reviews: https://www.infoworld.com/article/3141605/artificial-intelligence/ review-spark-lights-up-machine-learning.html

Website: https://spark.apache.org/mllib/

Apache FlinkML

Apache Flink provides a platform used for a streaming engine for the data flow that includes communication, data-distribution and fault-tolerance for data streams over distributed computations. It is an open source framework, which used for batch data processing and distributed stream. Apache Flink has a rich library, which used to solve any complicated problems. It is also capable of resolving any complex issues related to analysis. It is an open source stream processor, which helps in your business environment by quickly react to the most recent changes. Flink is faster because of its streaming architecture, and it processes data faster than Spark. It increases the performance of the job by instructing to only process part of data that have changed.

Type: Framework

Founded by Apache Software in August 26, 2014.

GUI Supported: No

Features:

- Fault tolerance and streaming first
- Scalable and high performance
- It supports *Java, Scala, Python and R.*

Comparison:

- Apache Flink provides a single runtime for the streaming and batch processing as compared to other streaming software, which don't provide such features.
- It supports controlled cyclic dependency graph in run time and it provides automatic memory management.
- It iterates data by using its streaming architecture.

Access/Source: Open Access/Open Source

License: Apache License v2.0

Language: Java, Scala.

API-Supported:https://ci.apache.org/projects/flink/flink-docs-stable/dev/api_concepts.html

Total Users/Products: https://flink.apache.org/community.html#project-wiki

Platform: Windows, Linux.

Community: https://flink.apache.org/community.html

Videos: https://www.youtube.com/watch?v=aWJ7CkSKMpQ

Reviews: https://flink.apache.org/news/2017/12/21/2017-year-in-review.html

Website: https://flink.apache.org/

Mahout

Apache Mahout, a project of the apache software foundation, offers a variety of premade algorithms focused primarily in the areas of collaborative filtering, clustering and classification. Mahout also offers Java-based libraries for common math operations focused on linear algebra and statistics. This framework is a work in progress, the number of algorithms implemented has grown fast, but many are still missing.

While mahout's core algorithms for clustering, classification and batch based collaborative filtering are implemented on top of apache hadoop, contributions to hadoop-based implementations aren't restricted. Apache mahout envisions transforming big data into big information faster.

Apache Hadoop

Apache Hadoop is an open-source framework developed for distributed computing of data across large clusters. The hardware dependency is minimized by exception detection and handling at the application layer, maintaining the high availability. Various distribution platforms like cloudera and IBM open platform have been built on top of hadoop. All the tech giants use from amazon to yahoo use hadoop in one way or another.

The core components of hadoop framework are the cluster, host machines, YARN-'The framework responsible for management of resources and job assignment', and map reduce. YARN based implementation of the map reduce model for parallel data processing and hadoop distributed file system. A high bandwidth file system to provide data to applications. This architecture along with data locality can surpass a supercomputer in processing large data sets.

Hadoop is released under the apache license 2.0. Primarily written in Java, many third-party programming libraries have surfaced to integrate hadoop with "hadoop streaming". Its use is specifically applicable to big data, and is largely used in financial, retail and healthcare sectors for both academic and commercial purposes.

LibFeatures

LibFeature is a software library for building feature vectors from structured data. It is particularly helpful for machine learning and data mining researchers who are using general-purpose algorithms to a broad variety of data sets where the data organized in different formats. Using this software, you can write a description of the features in high-level to extract from the input memory array or input files, and it dynamically computes the feature vectors you want on the fly. It dramatically simplifies the task of removing feature vectors from raw data. It can be executed on many feature vectors in parallel, resulting in performance comparable to efficient C code.

Cellerator

Cellerator is a mathematica package, which provides an automated equation generation designed to facilitate biological modelling. The implementation based on the idea of a hierarchy of canonical forms that represent biological processes at multiple levels of detail. Each level of the regime there are two classes of canonical forms can identify the canonical input form, that is used to provide information to the program, and the canonical output form that is produced by the simulator.

Anomaly Detection Tools

Prelert

Prelert is an anomaly detection software, which is used for automatically learning the behavior of applications and predictive analysis. Prelert software is a splunk app that can improve its feature into anomaly detection through machine learning process. It is a layer of highly advanced predictive analytics software, that easily integrates with and turbocharges your existing administration tools. It enables genuinely proactive management by automatically learning the typical performance of your application and supporting environment and alerting you to potential difficulties as they develop.

In prelert software, performance administration tools are provided that can automatically separate the causality of application interruptions in real-time. It provides the solutions that include a business service management solution that isolates application performance abnormality and root-causality in real-time. The software implements a feature called as causality analysis affirmation process, in which it takes existing data, and the data has automatically recognized the events of causality, such as trend, usage service on management telemetry, and streaming event, which lead to application errors. It offers its solutions for market data systems, web-based applications, internal network applications, and online banking systems environments.

Type: Software

Released in 2009.

GUI Supported: Yes

Features:

- Detecting advanced security threat activities and anomalies in log data,
- Discovering hidden fraud patterns in highly sensitive data,
- Identifying anomalous systems or metrics and their root causes across IT systems,
- Linking together complex series of events in data to expose early warning signals,
- Automatically pinpointing where and why critical system outages are occurring,
- Detecting unexpected drops in transactional activity, and much more.

Comparison:

- Automatically learning the behavior of application and predictive analytics as compared with splunk, which don't provide automatically learning.
- Highly advance predictive analytics features as compared to others softwares.
- Finding root causality in real time and web based application and internal network application.

Access/Source: Closed Access/Closed Source

License: Paid

API-Supported:
http://www.prelert.com/docs/products/latest/engine_api_guide/index.html

Platform: Windows, Linux, Mac

Community: https://www.elastic.co/prelert/support

Videos: https://www.youtube.com/watch?v=rgISOzwSbjE

Reviews: https://www.itqlick.com/prelert

Website: https://www.elastic.co/blog/welcome-prelert-to-the-elastic-team

Anomaly Detection

Anomaly detection is open-source or easy to use package for R platform and it is used to identify exceptions, which are strong, from a mathematical viewpoint, in the nearness in regularly, and an underlying course. Anomaly detection is an anomaly detection package, which is a self-learning predictive analytics with machine intelligence assistance, identify both normal and abnormal machine behavior. This package is worked on extremely advanced pattern recognition algorithms and provides detailed diagnostic data, and it identifies developing issues and allowing professionals to bypass problems or diagnose them. The package in the anomaly detection can be used in a broad diversity of settings. For example, after a new package release, it can identify anomalies in system metrics, A/B test post by user engagement, or for problems in social science, financial engineering, and political. The anomaly detection package is carried out on periodical support. For example, at intervals, one may be interested in discovering whether there was an anomaly yesterday for this, we support a flag only last whereby one can subset the irregularities that occurred during the last day or last hour.

Type: Package

Released in 2015

GUI Supported: No

Access/Source: Open Access/Open Source

License: GNU 3.0

Language: R

API Supported: No

Platform: Windows

Community: No

Videos: https://www.youtube.com/watch?v=Ebdp5Ao1o9o

Reviews: No

Website: https://github.com/twitter/AnomalyDetection

NuPic

NuPIC or the Numenta Platform for intelligent computing, is an HTM(Hierarchical Temporal Memory) implementation created by Numenta and based on a theory of neocortex, a part of brain involving high-level thinking. It was open-sourced in June 2013. NuPIC is highly efficient at anomaly detection in streaming data. It even has the numenta anomaly benchmark (NAB) with labeled data files and a scoring mechanism to compare real-time anomaly detection-algorithms.

Anodot

Anodot is a behavioral analytics system, which identifies outliers in the data. The detection is automatic, so that the blind spots can be identified in real-time. Insights from multiple anomalies can be extracted without alerting the user. The result is then delivered with a possible remedy. Interactive visualizations are included n the report to comprehend the insights easily.

High-volume data processing and adaptive nature of the algorithms makes Anodot a popular choice for anomaly detection. The pricing model for Anodot is quote based and it is available for all types of enterprises. The app is available cross-platform as is accessible from the different web browsers.

Loom Systems

Loom Systems provide anomaly detection software, which can automatically read the logs and identify behavioral anomalies. Thus, there is no need for data pre-processing. Even unstructured logs can be read, as loom automatically structures it. These anomalies are then reported along with interactive visualizations. Third party integration with Slack and other tools is also available. It is best known for event correlation and built-in recommendations. Loom Systems has a subscription-based model.

Interana

Interana is a behavioral analytics platform with an ability to analyze conventional business metrics in real-time. It is a solution highly scalable and capable of exploring and monitoring huge datasets. It is comparatively faster than other platforms and is ideal for processing event-based data. The UI is designed with an ease-of -access even to the non-technical users. Questioning for deriving insights is done through type heads and dropdowns. The business model is quote based and a free demo can be requested by an enterprise.

Machine Learning

Encog

It is one of the open source machine-learning framework for .NET or Java. Encog provides an advanced machine-learning framework that supports a variety of advanced machine learning algorithms, as well as it helps classes to process data and normalize the data. Machine learning algorithms such as neural networks, bayesian networks, genetic programming, svm, hidden markov models and genetic algorithms supported in this framework. Most of the training algorithms are compare well to multicore hardware, and it is multi-threaded. A graphical user interface is also provided to help model and train machine learning algorithms based on the workbench.

Type: Framework.

Founded by Jeff Heaton in 2008.

GUI Supported: Yes.

Features:

- Support of machine learning algorithms such as support vector machines.
- Support of artificial neural networks.
- Support of genetic programming, bayesian networks, hidden markov models, genetic programming and genetic algorithms.

Comparison:

- It supports gui for easy to use environment as compared to other frameworks like Pybrain which doesn't support gui.
- It supports more languages like Java, .NET, CS as compared to other frameworks.

Access/Source: Closed Access/Open Source.

License: Apache 2.0

Language: Java, .NET.

API-Supported:http://heatonresearch-site.s3-website-us-east-1.amazonaws.com/ javadoc/encog-3.3/index.html

Total Users/Projects: https://github.com/encog

Platform: Windows, Linux.

Community: No

Videos: https://www.youtube.com/watch?v=224EHp6mhNo

Reviews:http://www.discoversdk.com/products/encog-machine-learning-framework#/overview

Website: https://www.heatonresearch.com/encog/

PyBrain

PyBrain is a modular and open-source machine-learning library for python. Its purpose is to give flexible, simple-to-use yet still robust algorithms for machine learning tasks and a type of predefined environments to test and compare your algorithms. PyBrain can be trained and manipulated with almost all of the algorithms and has importance on network architectures. Environments in PyBrain are scenarios or test cases, in which an agent can prepare, or an algorithm can test. Environments share a standard interface, thus making it very easy to switch agents.

Type: Library

Released in July 14, 2009.

GUI Supported: No

Features:

- Open source python library.
- Easy to install and use and support for neural networks algorithm
- Simple and flexible, the algorithms are easy to learn and implement.
- Good support for Neural Networks, you can deal with neural networks at every level.

Comparison:

- This library has state of art neural network algorithms with a very powerful API that enables you to work with every level of building a neural network with ease as compared to other libraries, which doesn't provide api support.
- PyBrain has simple API, which can be understood by even a newbie as compared to other python libraries.

Access/Source: Closed Access/Open Source

License: BSD 3-Clause

Language: Python

API Supported: http://pybrain.org/docs/

Platform: Windows, Linux

Community: No

Videos: https://www.youtube.com/watch?v=fEM7YDNonSE

Reviews: https://www.g2crowd.com/products/pybrain/reviews

Website: http://pybrain.org/

Dlib

Dlib is a library which is a general-purpose cross-platform open source library written in C++ programming language. Dlib is a new C++ toolkit, which is used for creating complicated software and this library contains machine learning, computer vision, image processing and linear algebra tools and algorithms to solve real-world problems. This library is used in a wide range of areas including robotics, vision software, cameras, mobile phones, and sizeable high-performance computing conditions. Ideas profoundly influence its design from the design by contract and component-based software engineering.

In C++ application, most of the library is just header files that you can include. Moreover, if you are a python programmer, no worries, it has python API as well ready to use. The new version of Dlib is out that is 19.0 and it has many new features, some of them like the new elastic net and quadratic program solvers. It also supports APIs for machine learning, deep learning, computer vision and other languages for use. If you are a professional software engineer who is working on embedded computer vision projects, and you are probably working in C++, then using those tools in these kinds of applications can be very useful.

Type: Library

Founded by Davis E King in 2002.

GUI Supported: No

Features:

- Contains machine learning algorithms and tools in order of creating complex software in C++ for solving real world problems.
- Provides complete and precise documentation for every class and function.
- High quality portable code and graphical model interface
- Elastic net and quadratic program solver

Comparison:

- It support machine learning, computer vision, deep learning algorithms as compared to opencv that don't support these algorithms.

- It support python API for having lot of features for image processing as compared to opencv.

Access/Source: Open Access/Open Source

License: Boost

Language: C++, Python, C#.

API Supported: http://dlib.net/api.html

Platform: Windows, Linux

Community: https://sourceforge.net/p/dclib/discussion/

Videos: https://www.youtube.com/watch?v=135MJJNwlO0

Reviews: https://sourceforge.net/projects/dclib/reviews/

Website: http://dlib.net/

Ai-one

Ai-one is a software development tools that enable developers/programmers to develop machine learning applications and artificial intelligence into software programs. It provides API for creating learning machines information spaces for processing and collecting data and information. Some of the tools are NDK, which is a toolbox for a database language. Topic-mapper, which is a solution for processing language in the form of text or data. Topic-mapper SDK, a solution allows developers to build text applications that deliver learning capabilities for semantic discovery, data collaboration, sentiment review, and classification.

Oryx 2

Oryx 2 is built on apache spark and apache kafka, specializing in real-time and large-scale machine learning. It comprises of both packages and end-to-end applications. The software has a three-tier architecture consisting of a generic lambda, specialization with ML abstractions, and an end-to-end implementation. Oryx is only applicable to a limited range of applications, such as collaborative filtering, clustering and classification based on random decision forests.

Accord.Net

Completely written in C#, Accord.Net is a .Net framework used for scientific computing. It was originally created to extend AForge.Net, but AForge.Net joined it instead. Consisting of a large development team, the framework has been used in multiple scientific publications and featured in many books.

Accord.Net was published under the GNU lesser public license. It is available for building production-grade applications such, even commercially. The primary applications of the framework are computer-vision, signal processing and statistics.

Shogun

If you want an open-source library written in a low-level language with incorporation of various algorithms and data structures, then Shogun is for you. Implemented in C++, it offers interfaces for Octave, Python, Lua, R, Java, and Ruby and C # using SWIG. It is released under the GNU General Public License.

Shogun primarily focuses on kernel machines such as support vector machines (SVMs) for classification and regression problems. Apart from offering a full implementation of Hidden Markov models, the library has community-based development with contributions to the core package and is used for education and research.

Scikit-learn

Another Python-based free machine-learning framework is scikit-learn. It comprises of variety of classification, clustering and regression algorithms including support vector machines, k-means and DBSCAN, random forests.

Scikit-learn has some core algorithms written in python to improve throughput. While still under active development, Scikit-learn was initially developed as a Google Summer of code project by David Cournapeau in 2007. Later, INIRA got involved, and the first public release was made in 2010 under BSD license. Apart from being reusable in many contexts, it is built on scientific and computational libraries NumPy and SciPy.

LIME

Understanding the predictions of a model holds a significant value in transforming it to a trustworthy model. LIME is one such tool to explain the workings of machine learning classifiers. Any black box classifier's behavior, which can be linearly approximated with it, if the classifier has two or more classes with raw text or numpy arrays as inputs. Currently, only individual predictions for text classifiers or classifiers acting on tables are supported.

Mlpack

A high-performance machine-learning library is mlpack, which puts the emphasis on speed and ease-of-use. It is written in C++ that built on top of Armadillo library and distributed under the BSD license. It can be used to develop both open-source and commercial software. The project is openly supported by georgia institute of technology and has a broad community of supporters all across the world.

The aim of creating mlpack was to make it easier for novices to start machine learning by using a simple API, and also utilizing the features of C++ to provide high throughput and consistency. It has been benchmarked with an automatic benchmarking system developed by Marcus Edel as a part of Google Summer of Code.

Yooreeka

This project started from a book called "Algorithms of the Intelligent Web" in 2009 and is now used for data mining, soft computing and machine learning. Yooreeka is written completely in Java and was initially released under LGPL license, but later versions moved to apache due to the restrictions. Mostly, clustering, classifications, search and recommendation algorithms are incorporated.

Both data mining and mathematical analysis can be done through Yooreeka. While it's a cross-platform library, it needs java pre-installed to operate.

VELES

VELES is a distributed platform, which provides machine learning and data processing services for a user. Znicz is a submodule of veles, and veles comes from samsung. It is written mostly in python, and it can run in an ipython notebook. It is bundled with mastodon, a subproject that integrates it with any java application. It provides a load balancer between java nodes and veles slaves, and it manages data parallel and model parallel or execution of models with the help of such easily allowing architecture.

DMTK

Distributed Machine Learning Toolkit is an open-source machine-learning toolkit that goes by the name DMTK. The DMTK kit contains a framework, which provides a training model on multiple servers and also a topic-model and a word-embedding algorithm for nlp. It provides a framework that supports unified interface for data parallelization, and big model storage for the hybrid data structure, model scheduling for significant model training, and high training efficiency automatic pipelining. DMTK framework provides a platform by which machine learning researchers can build their distributed algorithms based on machine learning with small modifications to their existing machine algorithms.

Amazon Machine Learning

Amazon machine learning provides a platform that enables developing a predictive application by applying algorithms and mathematical models based on the user's data. Amazon machine learning reads data through amazon s3, redshift and RDS, and then visualizes the data within the amazon machine learning API and the AWS management console. This information can be imported or exported to other AWS services via s3 buckets. It provides a straightforward graphical user interface to create machine-learning models from the data, which is stored in amazon's s3, and RDS and use these models for predictions by using amazon ML and wizards.

Speech & Language

Narrative Science

Narrative science provides the platform which teaching data never like before; with this technology, it interprets your data, and then transforms it into insightful, natural language narratives at different speed and scale. It turns your data into an asset, which is actionable, powerful, and you can use to make better choices to improve communications with customers and empower your employees. It based on a unique concept in the mind that it generates automated summaries using data. It uses high-level natural language processing to create reports to operate as a data story-telling device. It is something similar to a consulting record. Narrative science currently used in financial sectors, insurance sectors, government and e-commerce domains. Some of its customers are forbes, deloitte, american century Investments, Pay Scale, MasterCard, etc.

Type: Speech

Founded by Kristian Hammond, Larry Birnbaum, Stuart Frankel in 2010.

GUI Supported: Yes

Features:

- Seamlessly integrated within power BI.
- Dynamic narratives that update in real-time as a user filters the analysis.
- Narratives for line charts, bar charts, pie charts, tree maps, histograms, and scatterplots.
- Options to personalize narratives by providing additional context including data characterization and analytic packages to run.
- Ability to customize the format and length of narratives.
- Automatically receive perceptive and accurate insights.
- Spend less time interpreting, simply read.

Comparison:

- Automatically access data and generates intelligent narratives as compared to mallet, which doesn't provide automatically access data feature.
- Personalized narratives and machine scale integration.

- Generates automated summaries using data, which is used for easy understanding of the natural language.

Access/Source: Closed Access/ Closed Source

License: Paid

API Supported: Yes

Platform: Add-on, Extension

Community: https://narrativescience.com/Resources

Videos: https://www.youtube.com/watch?v=i6jx42dMyxQ

Reviews:https://www.g2crowd.com/products/narrative-science-quill-engage/ reviews

Website: https://narrativescience.com/

Mallet

Machine learning for language toolkit or mallet is a java-based integrated package useful for natural language processing, document classification and other machine learning applications. There are sophisticated tools for each model, and a wide variety of algorithms that can be assessed, and implemented. Mallet is freely available under the common public license. It provides the ability to analyze unlabeled text and optimize numerical representations.

AaltoASR

AaltoASR, standing for aalto automatic speech recognition tools is a speech recognition toolkit developed by Aalto University. The prime components are Aku- 'the acoustic modeling toolkit', Aalto Decoder-'a C++ library that recognizes speech based on probabilistic computations from Aku' and PyRecTool-'a Python based software that automatically recognizes speech using ASR tools'. It is released with a Modified BSD license.

Barista

Developed by USC-SAIL, Barista is a customizable concurrent speech-processing framework. It is designed based on the kaldi toolkit and libcppa library. The structure is based on the actor model, where an individual actor is independent, and it can communicate with other actors. It can be implemented on both clusters as well as multi-processor CPUs.

Barista is an open-source framework, and release under the apache 2.0 license. It aims to use the parallel processing resources available for speech recognition. Its early release is available on GitHub, but it's still under development.

OpenNLP

Apache's open natural language processing toolkit is a popular choice among users because of its extensibility and since it provides most of the common NLP APIs, through which even more advanced speech recognition tasks can be built. The project is written in java and is available as a command-line tool but can also be integrated with other products. It has an easy-to-use Training API. OpenNLP has two primary cons which are the code is not frequently updated and the documentation is scrambled. It is released under the apache license 2.0 and can be used for commercial projects freely.

OxLM

The oxford neural language modelling toolkit attempts to resolve the data sparsity issue by learning distributed representations of words in a continuous vector space. It follows the log-bilinear language model, which is based on similarity scores and the application of gradient descent. The framework has been written in C++ by which the oxford computational linguistics group. The code is freely available on GitHub and can be integrated with Cmake. It is also compatible with cdec and moses.

Bavieca

The Bavieca automatic speech recognition toolkit is written in C++ and emphasizes on reliability and scalability. It is open-sourced and distributed under the apache license 2.0. The company developed the software, named as boulder language technologies for research purposes. It supports acoustic modeling based on continuous density hidden markov models (CD-HMMs). Bavieca has shown viable results in WSJ command line tools are available in the toolkit in addition to the Bavieca c++ api. A Java API is also available to easily integrate Bavieca with java applications.

Data Mining

OpenNN

OpenNN is a c++ library, which involves implementation of neural networks. For supervised learning, any number of layers of non-linear processing units can be implemented by the software. This deep architecture allows accounting universal approximation properties in the design of neural networks. Additionally, in order to increase efficiency, multiprocessing is allowed by means of openNN.

OpenNN contains data mining algorithms as a bundle of functions, which can be embedded in other tools, using an interface, for the integration of the predictive analytics tasks. Neural designer is built on top of OpenNN. In this regard, visualization tools can support some functions, but a graphical interface is missing.

Type: Software

Released in 2003

GUI Supported: No

Access: Free

License: LGPL

Language: C++

API Supported: No

Platform: Linux, Windows, mac

Video: https://www.youtube.com/watch?v=U_-JThd-_uc

Website: http://www.opennn.net/

Repository: https://github.com/Artelnics/OpenNN

WEKA

If you require a complete Data mining suite for machine learning, waikato environment for knowledge analysis (Weka) can be your go-to choice. It is an open source software developed at the University of Waikato, New Zealand, and released under GNU general public license. Written in java, Weka comprises of learning algorithms for data mining tasks and graphical user interfaces along with visualization tools. It is portable, multi-purpose, and easy to use and has online courses available for learners.

Originally, weka was designed to analyze data from agricultural domains. However, the recent versions are used in numerous fields, particularly for educational purposes and research. It is currently not adept at multi-relational learning.

Type: Software

Released in 2003

GUI Supported: Yes

Access: Free

License: GNU General Public License

Language: Java

API Supported: Yes

Platform: Linux, Windows, OS X

Community: https://www.cs.waikato.ac.nz/ml/weka/help.html

Video: https://www.youtube.com/watch?v=m7kpIBGEdkI

Review: https://www.capterra.com/p/171134/Weka/

Website: http://www.cs.waikato.ac.nz/~ml/weka

Repository: https://svn.cms.waikato.ac.nz/svn/weka/

Massive Online Analysis (MOA)

Another popular open-source framework is MOA, with which experiments can be developed and run specific to data stream mining. Written in java, it is consist of a set of learners and stream generators. A variety of algorithms designed for large-scale machine learning is constituted in MOA with a support for bi-directional interaction with WEKA.

MOA is released under GNU GPL license. It offers both online and offline methods for evaluation and its objective is to assess a benchmark for the stream mining community.

Type: Software

Released in 2010

GUI Supported: Yes

Access: Free

License: GNU General Public License

API Supported: Yes

Platform: Linux, Windows, mac

Community: https://moa.cms.waikato.ac.nz/people/contact/

Video: https://www.youtube.com/watch?v=9-WZmdu4cPE

Website: http://moa.cms.waikato.ac.nz/

Repository: https://github.com/waikato/moa

ELKI

Environment for developing KDD-applications supported by index-structures (ELKI) framework is written in java with currently incorporated algorithms being clustering, outlier detection and database indexes. Initially aimed at being used for research and study, it is now used widely under the gnu affero license. It aims towards the progression of advanced data mining techniques. ELKI's application requires thorough study of algorithms and manual integration with business intelligence applications. Furthermore, commercial usage is inhibited due to the copy-left license.

Type: Software

Released in 2003

GUI Supported: No

Access: Free

License: AGPL v3

API Supported: Yes

Platform: Linux, Windows, mac

Video: https://www.youtube.com/watch?v=jj6-0cvcNEA

Review: https://www.predictiveanalyticstoday.com/elki/

Website: https://elki-project.github.io/

Repository: https://github.com/elki-project/elki

Mlpy

Machine learning python or Mlpy is open-source library providing advanced machine learning methods for solving both supervised and unsupervised problems. It is designed on top of GNU scientific library and NumPy/SciPy with an objective to maintain a balance between various features such as modularity, reproducibility and efficiency. Released under the GNU license, Mlpy is free to use and extend commercially. Most high-level functions and classes are provided with the ability to perform numerous statistical analyses.

Oracle Data Mining

Oracle provides numerous data mining modules through it's relational database management system. The database environment incorporates both creation and deployment tools with a database kernel to store the objects. There is also a database vault to maximize security.

Available with a proprietary license, oracle offers a data miner gui for development of models. Initially introduced in 2002, oracle data mining succeeded the darwin data mining toolset. All the popular machine learning approaches and algorithms are incorporated, with no need to convert or move the data. Its powerful mining ability, real time management of applications along with the different management packs for data makes it a popular choice among businesses.

Orange

Another handy data mining and analysis tool is Orange. It allows interactive data mining through visual programming and different types of components termed as widgets. Over 100 widgets can be customized according to the user. It is a cross-platform software with initial release in 1997 under GNU license. Orange utilizes python-based libraries for numerical computation and has a canvas-like interface where interactive visualizations are performed.

Neon

Neon provides an open-source framework, which is used for analysis applications by developing custom data. It is created to visualize and integrate yours with your custom data. The neon structure can incorporate a variety of data analysis and visualization widgets in an everyday web environment to allow analysts to explore elaborate data sets. It provides the bridge between visualization and data stores widgets. Allows developers to create data exploration applications and workflows agnostic of other organization has input. Permits users to send a query to No-SQL databases using a SQL-like language.

Neural Designer

Neural Designer is a desktop application, which used for data mining, which uses neural networks, the central paradigm of machine learning. It is a professional application, which is used for exploring complex relationships, predicting actual trends from data through neural networks and recognizing unknown patterns. Some of the occasions where neural designer has used such as to reduce consumption and increase comfort in aircraft by flight data, in medical databases to make more reliable and less invasive diagnosis. This application has also used in physicochemical in which the sales data to optimize provisioning and to improve work quadrants, and the data has to enhance the quality of wines.

END OF THE BOOK

We have left out several specialized softwares and packages because we wanted this book to be a guidebook of tools for the majority of AI learners and practitioners. We would publish another book soon, which would be a bit more voluminous than this and serve the needs of even larger AI population.